For Melissa,
a rancher at heart!
Enjoy,
Barbara Van Cleve
September 15, 2000

The western expression "Hard Twist" refers to the oldtime Manila-hemp, tightly twisted lariat rope—*hard twisted.* The term also refers to a small, compact, physically strong person with resilience, rather like rawhide, which expands and stretches when wet or shrinks and tightens up when dry but almost never breaks.

HARD TWIST

WESTERN RANCH WOMEN

PHOTOGRAPHS BY
BARBARA VAN CLEVE

MUSEUM OF NEW MEXICO PRESS
SANTA FE

For my Mother,
Barbara,
with great respect, admiration, and love

Project editor: Mary Wachs
Design and production: David Skolkin
Typesetting: Set in Walbaum with Walbaum bold display by Wilsted & Taylor
Manufactured in Singapore
10 9 8 7 6 5 4 3 2 1

Library of Congress Cataloging-in-Publication Data available.
ISBN 0-89013-287-9 (CB) ; 0-89013-293-3

Museum of New Mexico Press
Post Office Box 2087
Santa Fe, New Mexico 87504

Foreword

The subject of Barbara Van Cleve's photographs is the women of the West. When Cartier Bresson was asked what told him that something should be photographed, he replied, "My general culture." This is exactly what informs Van Cleve's pictures: she is working from the middle of a world she knows like the back of her hand. The magic of art is that our instincts, whoever we are, confirm these truths, even in the absence of personal experience or information. I can't think that anyone, however far from this milieu, would be unmoved by the intimacy and perception of Barbara Van Cleve's West.

There is a kind of pent-up energy in these pictures. Partly, it is the charge of all good work, including photographs. But the subjects account for it too because they convey truth, insufficiently known but containing the velocity of recognition.

We have not sufficiently been made to feel the presence of women in the history of the West. In the catch-up that has ensued in recent years, one often had the sense of quota building, of revision from above, the usual assortment of tastemakers and social engineers from elsewhere. But with the most cursory look at Barbara Van Cleve's pictures one knows immediately and deeply that women were always the heart of this story. Barbara Van Cleve is not arguing with anyone: the truth of her vision is deeper than polemics. It is solider. She is not revising a story so much as completing one that was barely half told.

The senselessness one always felt at the chronicle of the West, not just the easily dismissed movies and pulp fiction but the official account that emerges from our most august regional historians and honored novelists, from our most entrenched academics, is undermined by the closeness and truthfulness of these photographs. The women who in their daily lives have built this distinctive culture have quietly been at this work for a very long time. They have always been here and it has always mattered. For those of us who love the West and make our lives in it, this news has the power of the best things that are saved for last.

—Thomas McGuane
McLeod, Montana

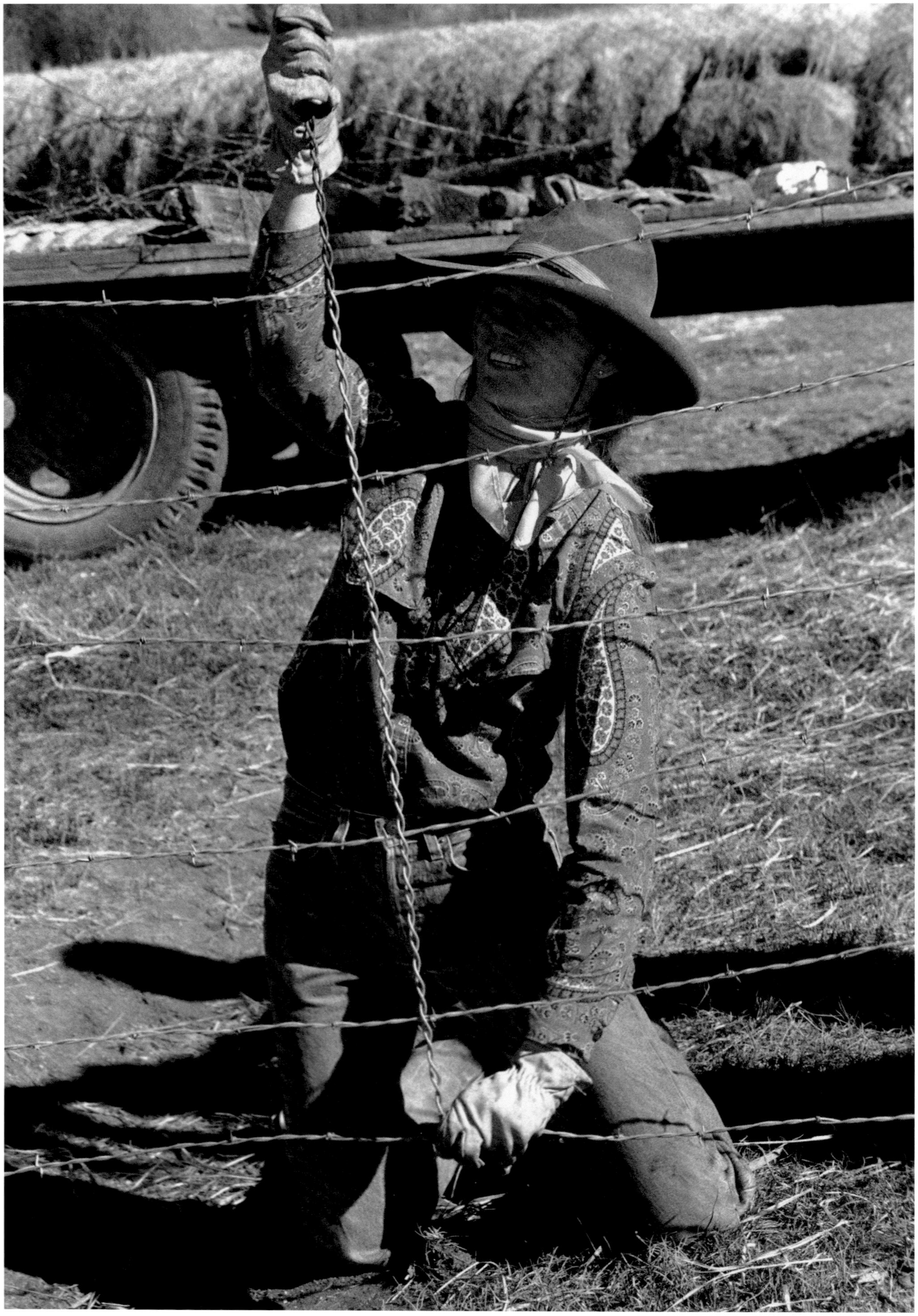

"TOOTIE" MITCHELL HANSEN, STAKE RANCH, ELKO, NEVADA
1986

Preface

It has taken too many years to get this book published, especially since fewer years were needed to travel to the ranches, photograph and interview these typical yet remarkable women. Editing our long, rambling, but interesting, conversations down to hundreds of words instead of thousands of words was downright painful. But as my editor so often reminded me, this is a photographic book, not documentary literature.

I am deeply appreciative to Michael Duty for his belief in this project and for his work with Byron Price, director of the National Cowboy Hall of Fame and Western Heritage Center in Oklahoma City, which will open the major exhibition of these photographs and is sponsoring the subsequent two-year traveling show. I am pleased that Michael curated the exhibition and honored that he wrote a perceptive and illuminating afterword for the book.

I am indebted to Mary Wachs, editorial director of the Museum of New Mexico Press, for her enthusiastic desire to publish this book, for her cheerful and perceptive editorial suggestions, for her total professionalism and great sense of humor. To David Skolkin, art director of the Museum of New Mexico Press, I am grateful for his creative and persistent vision about the book's design. He is a master in his profession. He, Mary, and I engaged in spirited discussions about which photographs to include or exclude and yet remain friends. Both of these easterners have learned a great deal about animal husbandry as it is practiced on western ranches. They may not feel easy about various practices, but now they understand the need for them!

To Deede Phillips, my good friend and assistant, who "rode shotgun" those thousands of miles crisscrossing the Rocky Mountain West, my heartfelt thanks. Her unfailingly cheerful assistance at all hours and her daily support were invaluable.

Karolyn Wayman, my personal assistant of more than five years, who brought order to my studio and record keeping through her Macintosh expertise, typed all the first drafts from the transcripts, many of which ran forty pages and more. She has my deep appreciation for her patience, endurance, and hard work in helping me put this project into shape for presentation.

Anne Senstad, a young Norwegian photographer and graduate of the Parson School of Design, just thought she was coming to Santa Fe, New Mexico, for a vacation. She spent over half a month assisting me in printing the final prints for this book. Her dedication to making the best reproduction quality prints was tremendous and equally appreciated.

My appreciation also to Marge Piepmeier for transcribing all the taped interviews; to Paul Reinwald for making countless and beautiful work prints; to Denise Kusel for her enthusiastic encouragement and coverage over the years; and to Ted Flicker for his friendship and wisdom.

And most of all, to those women ranchers and their families who extended their hospitality and shared their lives and personal opinions with me.

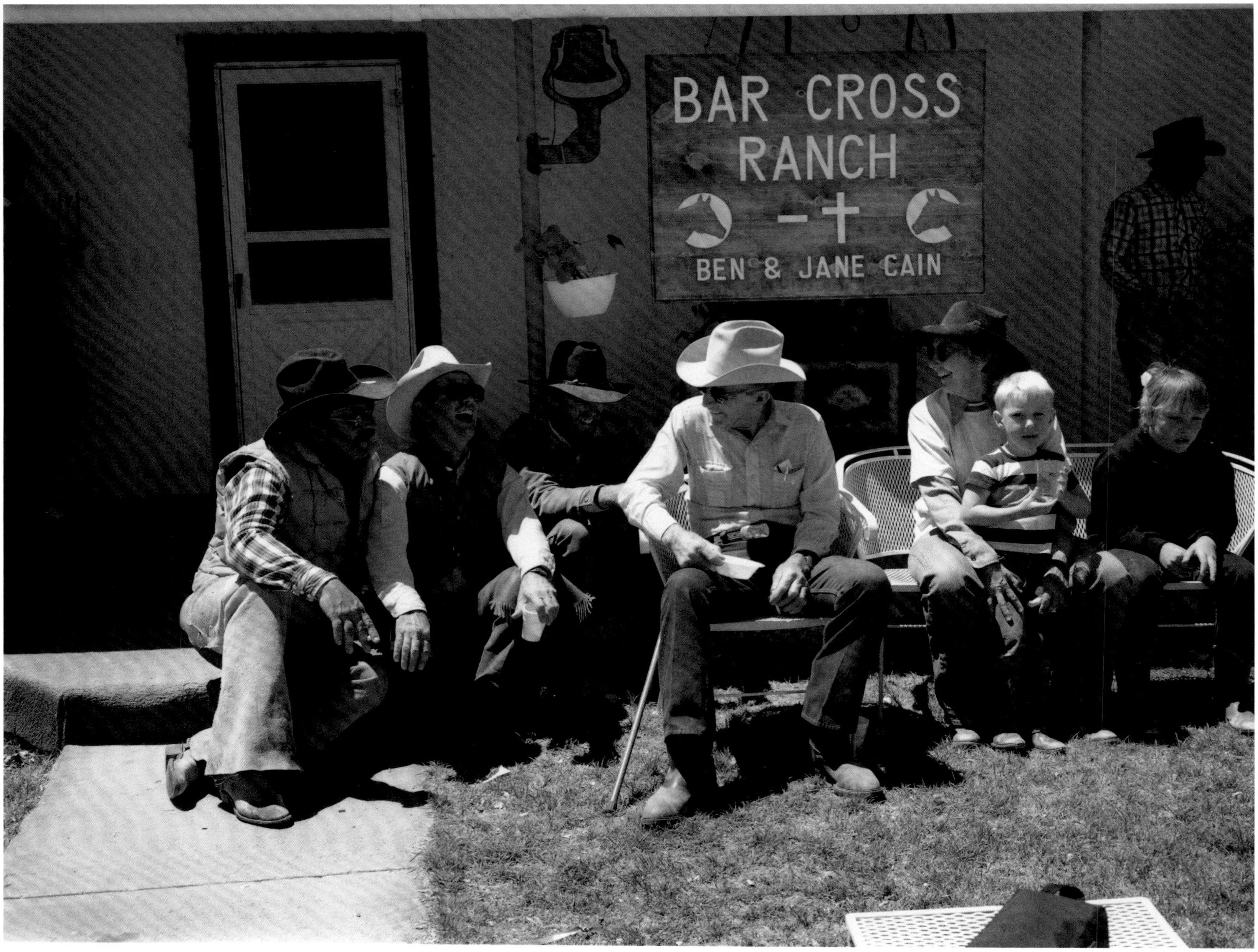

FAMILY AND NEIGHBORS GATHER ON CATTLE SHIPPING DAY,
BAR CROSS RANCH, ENGLE, NEW MEXICO
1988

I am far more grateful than these few words can adequately convey. Our verbal encounters were intimate, candid to be sure, and I am honored that they placed such trust in me. I truly hope that no one feels that I have betrayed the essence of what she was communicating. I had a lot of fun with each of you, exchanged stories, learned a lot about different ways of ranching, and came to appreciate the beauty of widely diverse landscapes. You each gave so much of yourself and your time. And I hope that together we have set the record straight about women ranchers!

Especially I want to thank the following women ranchers: Linda Mitchell Davis, Kim Davis Barmann, Julia Davis Stafford; Polly, Jeanne, and Dee Dee Dickinson; Gretchen Sammis and Ruby Gobble; Michele Carroccia, Cynthia May, Toni Schutte, Janie Carlson, Jane Glennie, Bobby Brooks Kramer, Jan Youren, Molly Flagg Knudtsen, Elladean Bittner, Melody Harding, Jewell McAdoo Lutich, Cynthia Rigden, Martha Jane Cain, Nina Robinson, Sharon Lusko, Celeste Sellers, Sandy Cook Rosenlund, Fern Sawyer, Karen Farr, "Tootie" Mitchell (Hansen), Evelyn Capps Walker, Claudia Capps, Rob Emery, Mary Hamilton, Karla Jo Camp, Ann Holland Daugherty, Barbara Saunders, Irene Walther, Eloise McQueary, Sharon Rhoades, Julie Early, Edna Mae Swain, Mildred Miller, Marlis Arenson Dunn, Doris Siebold, Marion Lloyd, Alisa Ogden Stell, Sally Schwartz, Ann Conlee, Mary and Alice Moore, Tanya Stevenson, Murial Noon, and Wanda Walker.

—Barbara Van Cleve
Santa Fe, New Mexico

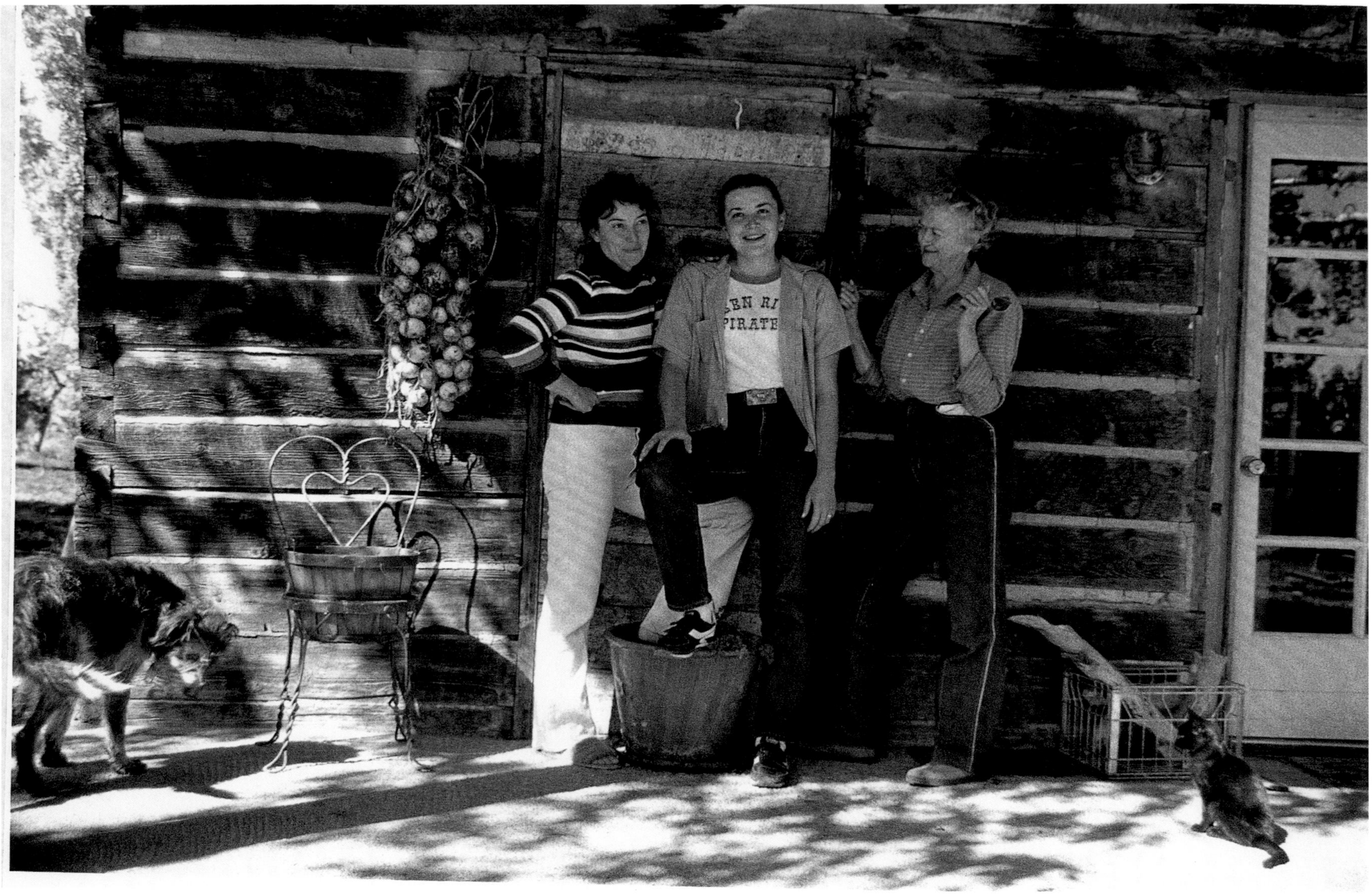

THREE GENERATIONS OF MORMON WOMEN ON THE FAIRVIEW RANCH, HANKSVILLE, UTAH. FROM RIGHT, NINA ROBINSON (GRANDMOTHER), CELESTE SELLERS (GRANDDAUGHTER), AND SHARON LUSKO (DAUGHTER)

1986

INTRODUCTION

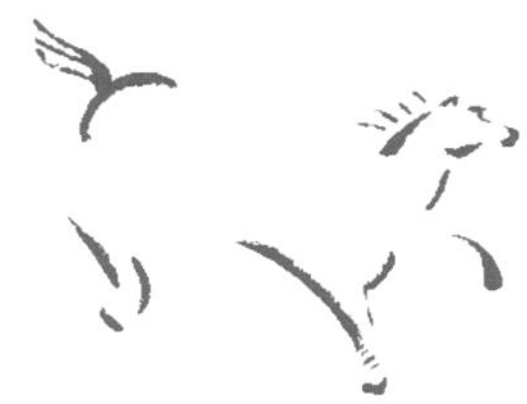

The role of women in the settlement of the West has been generally ignored until the last few years. Women were a necessary and powerful element in the migration to and settling of the West—in the "civilizing" of the frontier. Pioneering, like ranching, is usually viewed as the province of men, but the westward migration was definitely a family affair, as is ranching today. The women were absolutely essential, for not only did they perform their traditional tasks, which on the trail required a great deal of heavy work, but they also took over the men's work when necessary.

Most of the women did not want to make the journey west. The trip was marked by a lack of security and comfort as well as an uncertain future at best. Pioneers endured all kinds of weather with little or no shelter; the days were long and arduous; the women had to cook over open fires regardless of the weather; they had to care for the children who came with them as well as those they bore on the trail. On foot the road was long and dusty. Riding in a wagon (with no springs), perhaps a less comfortable option for a woman with child, was cold and jarring.

But these women were survivors. As a matter of fact, more men died on the trail than women, even with the difficulties and dangers of childbirth. For those of us born in the West, these pioneering women and men are our forebears, whose genes, passed down through the generations, collectively have resulted in a western spirit celebrated for its resourcefulness, toughness, and tenacity.

My ancestors were pioneers to be sure and an interesting group of people. My great-great-grandfather, Gen. Horatio Phillips Van Cleve, was posted to Fort Snelling (later to become the twin cities of Minneapolis and St. Paul) and married Charlotte Ouisconsin Clark. She is celebrated as the first white female born west of the Mississippi owing to her mother, Charlotte Oren Seymour, of Hartford, Connecticut, insisting on accompanying her husband, Maj. Nathan P. Clark (founder of Fort Snelling) into the wilderness. She gave birth to Charlotte only minutes after they had crossed the river. Given the middle name of Ouisconsin by the soldiers in her father's command, she spent her entire life at Fort Snelling. One of her legacies was the book *Threescore and Ten: Lifelong Memories of Ft. Snelling*, in which she described in detail the Indian attacks, plagues, killer winters, and the hardships of frontier life.

Charlotte's brother Egbert Malcolm Clark was born in 1817 at Fort Wayne, Indiana, where his father was an officer in the U.S. 5th Regiment of Infantry. He entered West Point at seventeen in the same class as W. T. Sherman but was expelled for duelling, whereupon he joined a relative, Sam Houston, in Galveston and was commissioned a lieutenant in the Texas Army, serving until 1836.

Chafing for the frontier, Egbert Malcolm Clark arranged for an appointment to one of the American Fur Company's stations on the upper Missouri, where he gained the respect of the Blackfoot Indians and married the daughter of a Peigan chief. In 1862

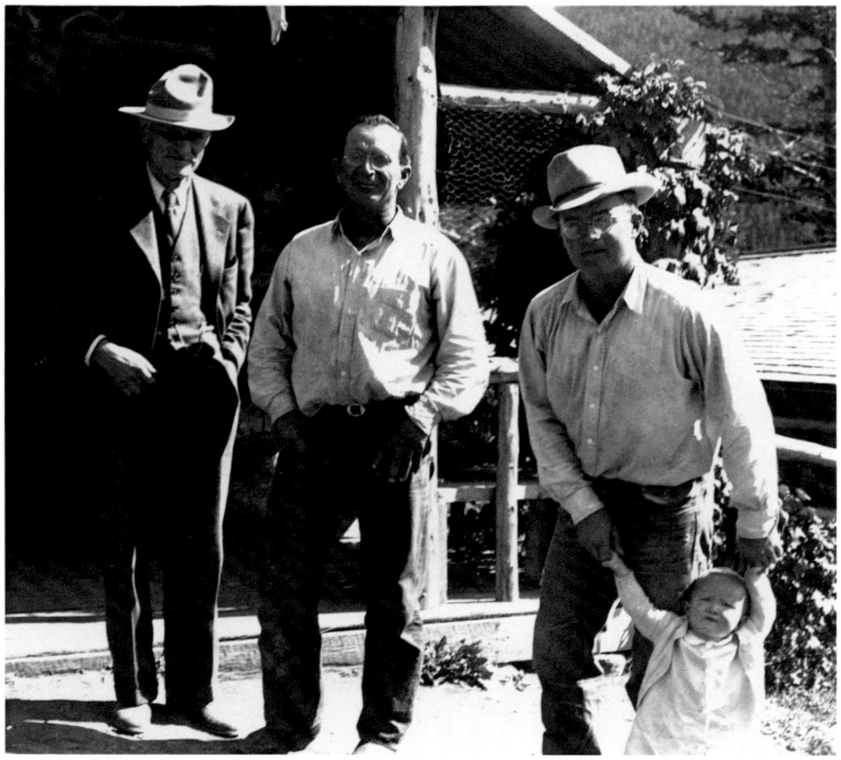

FOUR GENERATIONS (PATERNAL): PAUL L. VAN CLEVE, SR., PAUL L. VAN CLEVE, JR., PAUL L. "SPIKE" VAN CLEVE III, AND BARBIE, CA. 1937.

he left the fur business and settled down to ranch in the Little Prickly Pear Valley north of Helena, Montana Territory, where his treacherous murder in 1869 precipitated the infamous "Baker Massacre." Malcolm's sister, Mary Snelling Clark Lincoln, a devout Catholic of considerable means, insisted he send his daughters and sons to eastern Catholic schools. One of Mary's daughters, Florence, became Mother Angela, the famed pioneer Montana Ursuline nun. Malcolm's oldest daughter, Helen Piotopowaka Clark, became the first woman in Montana's state elective office—the Superintendent of Public Instruction.

One of General Van Cleve's sons, Paul Ledyard (my paternal great-grandfather), came to Montana Territory in 1878 at the request of his relatives, the family of Lt. John Crittenden, who was killed with General Custer at the Battle of the Little Big Horn, to dispose of property Crittenden had acquired in the territory. That duty done and finding he liked the country, Paul put his civil-engineering degree to work for the Northern Pacific Railroad, which was just then building its way across Montana Territory, as a surveyor and telegrapher. His wife, Alice Davis, came as far with him as the railhead of Coulson (later to become Billings), Montana Territory, where she operated a finishing school for young ladies while her husband was away surveying. Eventually he decided to settle in the area near the Norwegian settlement of Melville and go into ranching. Alice Davis continued to live in what was by then Billings and to conduct her finishing school while her husband erected the nucleus of what would become a thirty-six-room ranch house with indoor plumbing, steam heat, and piped acetylene gas lighting—all quite remarkable for a rural ranch house in the 1880s.

About this time, Alice got caught up in the railroad building fever that swept the country with the opening of the frontier. Unfortunately, she was at the tail end of the boom, and by the time she had invested thousands of dollars and lined up investors to build a railroad connecting the Northern Pacific at Big Timber to the Milwaukee at Harlowton, the bubble had burst nationwide. Her investors pulled out and the Van Cleves were deeply in debt.

Their only son, Paul Jr.—my paternal grandfather—who was attending Lawrenceville School in preparation to enter Princeton University, abruptly quit school, returned home, and assumed, at age sixteen, management of the ranch, paying off his parents' debts and going on to build another, much larger ranch. He married at age twenty Helen Perry, who was born and raised in Helena, Montana Territory, the daughter of Charles Theodore Perry of Switzerland and New York, a distant cousin of Commodore Perry. When he died, her father left her his rare art collection, which she sold to send her son, Spike (my father), to Berkshire School preparatory to entering Harvard University.

My foremothers on my mother's side of the family were equally strong and pioneering. My great-great-grandmother, Laura St. Clair Boal, abandoned the sophistication of St. Charles, Missouri, to marry Capt. Carroll James Atkins, a Missouri River steamboat pilot and owner, and eventually settle in Dakota Territory in 1884.

Her daughter, Alice Sinclair Atkins, my great-grandmother, was a tiny little ball of fire and ice, born in 1871. She came up the river from Missouri with her father and family, settling in North Dakota Territory. Later, she married Elisha Bartlett Page, who owned a lumber business in Oregon and was a banker, wheat elevator owner, and large-scale wheat farmer on the North Dakota prairie. Alice loved the prairie country of their farm and wrote and published poetry about her life on the land.

Alice had four children, two boys and two girls. Laura Williams Page, my maternal grandmother, born in 1893, was a ravishing red-haired beauty who met and was courted by Ingvald Lauritz Knudson, a second-generation Norwegian from Minnesota. They married in 1911 and settled in Harlowton, Montana, where he established the Farmers National Bank in 1918. They had three daughters, Loreen Sinclair, in 1912; Barbara Lenore (my mother), in 1914; and Carroll Virginia, in 1916. Ingvald died of stomach cancer, leaving Laura with three girls under the age of sixteen. Alone but undaunted, and vowing not to dip into the money he left for the girls' education, she started a real estate and insurance business while studying law (by correspondence) at night. Although she had only a high-school education when she got married, in three years she was able to pass the Montana bar with the highest score and was the second woman admitted to practice before the Montana Supreme Court. In 1956, she was admitted to practice law before the United States Supreme Court, the first woman from Montana to achieve this distinction.

Laura loved life and especially enjoyed opera, theater, travel, and entertaining family and friends. Upon retirement, she moved from Washington, D.C., to Big Timber, Montana, to be close to her daughter Barbara, my mother. It was during these retirement years that I really got to know my grandmother, whom I adored.

Barbara Knudson, a striking brunette with dignity and bearing, met my father, Paul "Spike" Van Cleve III, on a Girl Scout camp trip she was supervising. Backing out of a tent as he rode up on horseback, she heard him comment, "I sure hope the front end looks as good as the back end." So much for cautious introductions! He had just finished his first year at Harvard, but that summer his plans began to change. Their summer courtship was only briefly interrupted when Spike returned to Harvard, but before he could complete his second year, afraid that some other man would "beat his time" with Barbara, he returned to Montana and the family ranch. They were married September 26, 1934, in Minneapolis and eventually had one son and three daughters. I am the oldest of their children.

That fall, after a short honeymoon, they moved up to the only "winterized" log cabin in Big Timber Canyon. Back in 1922, Dad's parents, Paul and Helen Van Cleve, bought land in that canyon and hired a crew of Norwegian craftsmen to cut timber and build solid log cabins for the dude ranch operation they were adding to their regular ranch operation. This was one of the earliest working dude

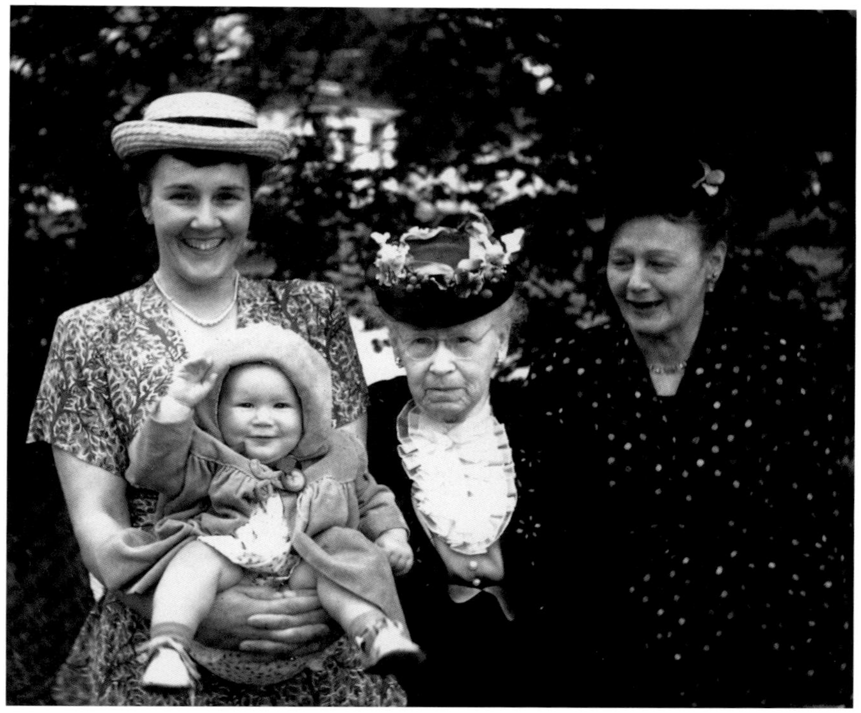

FOUR GENERATIONS (MATERNAL): FROM LEFT, BARBARA KNUDSON VAN CLEVE AND BABY CAROL, HER GRANDMOTHER, ALICE SINCLAIR ATKINS, AND MOTHER, LAURA PAGE KNUDSON, CA. 1948.

ranches in the West and is still family owned and operated today. My grandparents felt that not only were the guests interesting and even helpful, they also provided another source of income.

So Mother, a town girl who didn't even know how to cook, went with my father, a young "bucko," mountain man, and trapper (later to become a rancher and horseman), literally into the wilds. They had $13 to winter on that first year and they did! Mother became a terrific cook with Dad's tutelage. Today I am still convinced that she is the best cook of venison, no exceptions. Water had to be hauled up by hand from the creek down the hill from the cabin, so there was no question that they had "outside plumbing." An outhouse, drafty for ventilation, is an invigorating experience in Montana's below-zero winter weather. No one lingers reading the Sears and Roebuck catalog, that's for sure! During the three years they spent in the log cabin up Big Timber Canyon, totally isolated, Mother bore two children. She was all alone most of the days, waiting for Dad to come home from tending to his trap lines, hunting, cutting and hauling logs, alone. They heated the cabin with the wood cook stove but had no telephone, radio, or electricity. To communicate with anyone, Dad had to saddle a horse and ride some thirteen miles to his parents' ranch. It was truly an isolated life and very much like the lives of many of the early settlers of the West as well as the early lives of some of the women ranchers included here. With words, Dad paints a vivid picture of their life those first three years in "At the Head of the Creek."

The notion that the West was settled by men has persevered in the idea that ranching and cowboying are the exclusive domain of men. The same pioneer spirit and gutsiness that compelled those early women migrants to follow their husbands and eventually settle the West is embodied in today's women ranchers.

Women ranchers are "cut from the same cloth" as those early pioneering women. I remembered all the work Mother did, which I took for granted as I was growing up. She canned in the fall over the big wood-burning stove in the kitchen. It was hellishly hot in that kitchen! She churned cream by hand to make butter; she made bread by hand weekly and ironed our clothes with "sad irons," heavy cast-iron weights heated on top of the stove. During the summer she cooked three meals a day for the hay crew of five to seven men, in addition to all the other work. During the fall she had to cook for the threshing crew who cut the oat fields and thrashed the kernels out so we had oats for our livestock. We also had a garden which, after being planted, required hoeing regularly, never a favorite task. She rode with my dad to help gather and sort cattle throughout the year. In the winter, she drove the team, Bullet and Babe, while Dad fed the hay to the cows and their calves. During calving time, Mother always had the back room floor covered with several newly born calves, trying to keep them alive with house warmth, and the bathtub held a nearly frozen calf thawing out in tepid water.

Born and raised on a historic Montana ranch in an old ranching community, I became aware early in my life of who does double duty on a typical family ranch: the woman—mother, wife, widow, daughter, sister, or aunt. So many of the ranch women whom I knew growing up worked outside. They milked the cows, helped put up hay, helped with the fencing, drove the team to feed cattle in the winter, rode, helped work cattle, and sometimes even roped. They did these chores in addition to all the housework of cooking, cleaning, ironing, and, very often, bookkeeping. Ranch women thrive on the outdoors. They like the clean, fresh air and the sense of freedom and the sights and sounds of their vast western landscapes. They also work outside because they know that their work is essential to the success and survival of their family ranches. Contrary to public opinion, most ranchers are land poor and can only dream of being money rich. They depend upon the work of the entire family to survive and prosper.

Many women ranchers welcomed me onto their ranches and into their homes and allowed me to interview and photograph them during the eight years of this project beginning in 1986. The lifestyle of today's ranchers has been misrepresented in the media, and many were initially leery. However, because so many ranchers had read my dad's books or knew his name they felt that I could be trusted and that I wouldn't betray or misrepresent their position. I felt honored to be trusted so completely.

I drove thousands of miles to many ranches and always received a gracious welcome and warm hospitality. From the inception of this project I decided to photograph and interview selected women ranching in an area more or less designated as the Rocky Mountains from Canada to Mexico. And the women had to be on a real ranch, not just "forty acres and a dream," where they raised cattle at least. There is one sheep rancher (many of the big old-time ranches in some states were built on raising sheep, although now most of those have discontinued the sheep part of their operation and concentrate on cattle), and there are two ranches where they also raise a lot of horses or where they are restoring and cultivating habitat for wildlife and wildfowl.

I went back to various ranches at different times of the year so that I might photograph a wide variety of approaches to seasonal activities such as spring roundup and branding, irrigating, haying, fencing, shipping time, calving, and winter feeding. Every rancher does things a little differently, especially in the more southern part of the Rockies. This is high-desert country, with not much water unless they bring it up out of the ground, which is a pretty expensive proposition. I learned a great deal about different ways of ranching in the process of doing this project. Not wanting to "ride the grub line" (be a freeloader), when these ranchers were feeding me, providing a good bed and being so accommodating, I always offered to help do whatever work was needed. They usually took me up on my offer and outfitted me with a good horse (I had my own saddle, blanket, bridles, chaps, rope, and spurs), which enabled me also to photograph them from on horseback while they were out at work.

These women were wonderful, warm, funny, good natured, hard working, and generous to a fault. They have enriched my life greatly and I am privileged to count them as friends.

Lastly, I owe my dad so much gratitude for teaching me early on about seeing the beauty in the landscape, weather, and in the dirty, hard work required in ranching. At six years or so I was helping move cattle. I remember complaining to him because we hadn't eaten and it was 10 degrees, so I was cold and hungry. He said, "Red, I need you to help me here, so you might as well sit up and learn to see the beauty in it." I never forgot that. And that day in January he showed me, for the first time, a mirage down on the low country horizon and a frost haze with dancing particles of colored light. He taught me how to see!

But as time passes, so do family members. Life changes. In 1982, Dad died unexpectedly of a heart attack, in Mother's arms, at the ranch. During nearly forty-eight years of marriage I don't think they had spent more than twenty-four hours apart at any one time. Today, Mother continues to live an active life on the family ranch, and, at eighty-one, is still the beautiful woman Dad fell in love with so many years ago.

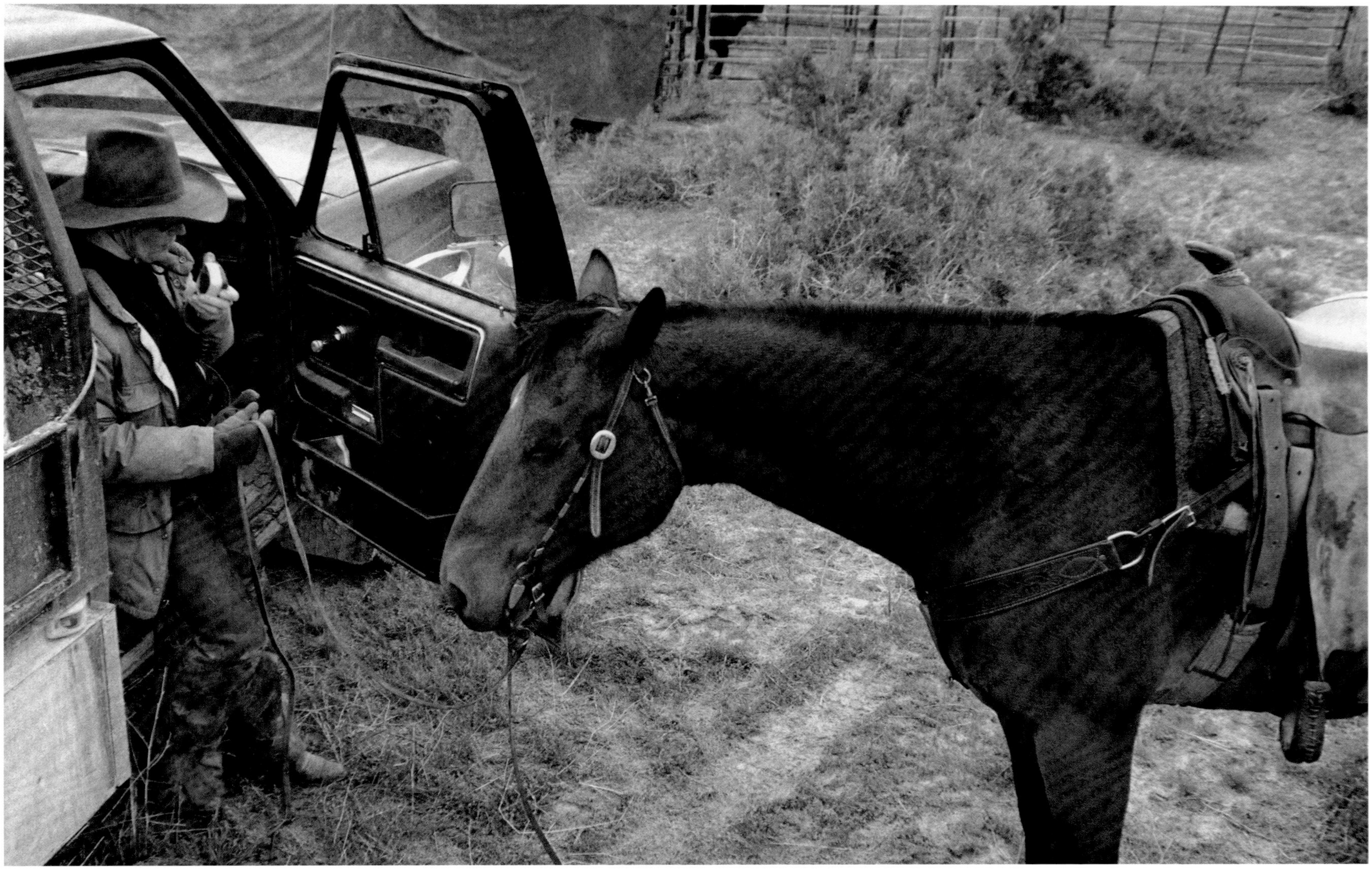

POLLY DICKINSON, VERMILLION RANCH, MAYBELL, COLORADO
1987

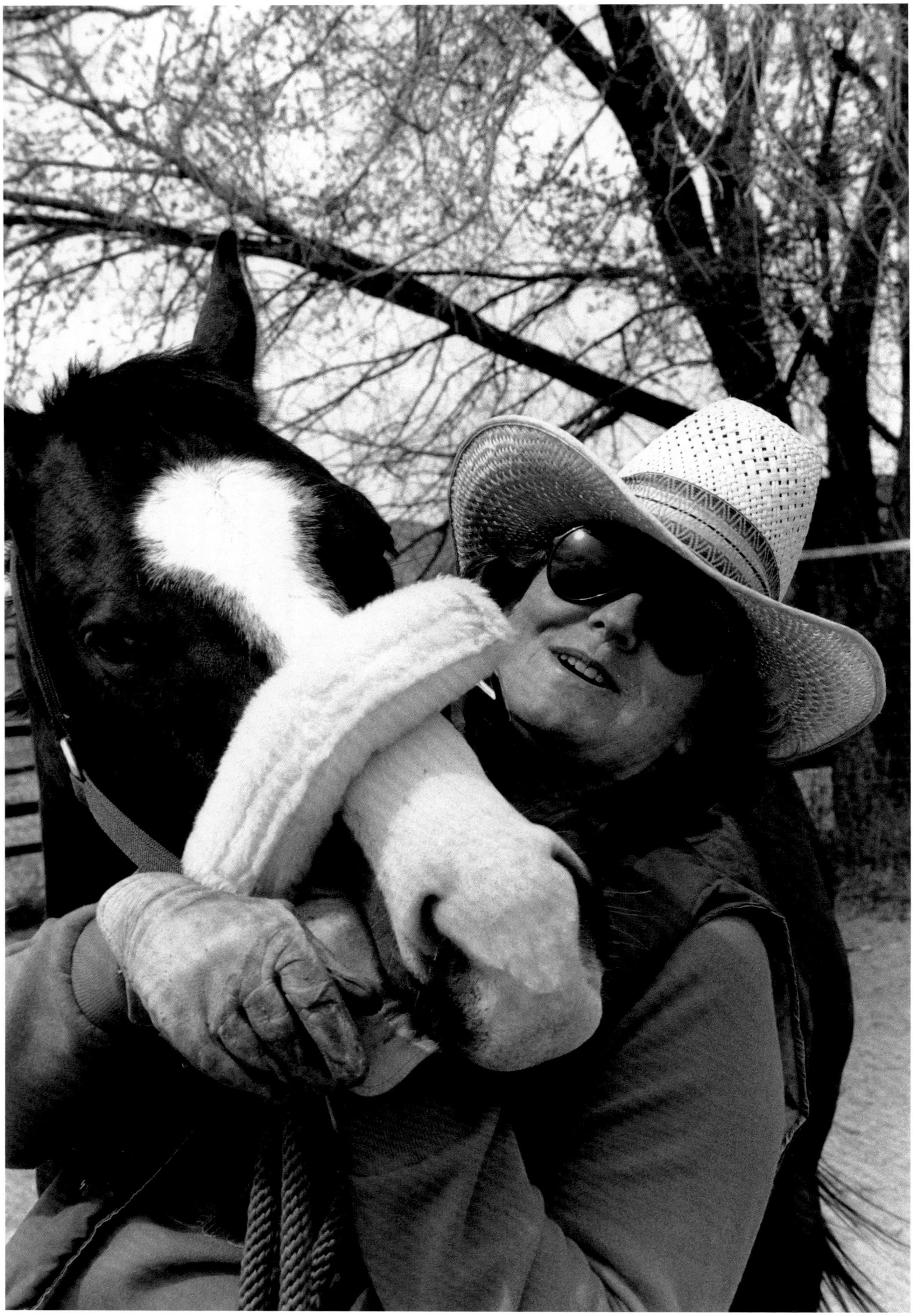

CYNTHIA RIGDEN, RIGDEN RANCH, KIRKLAND, ARIZONA
1987

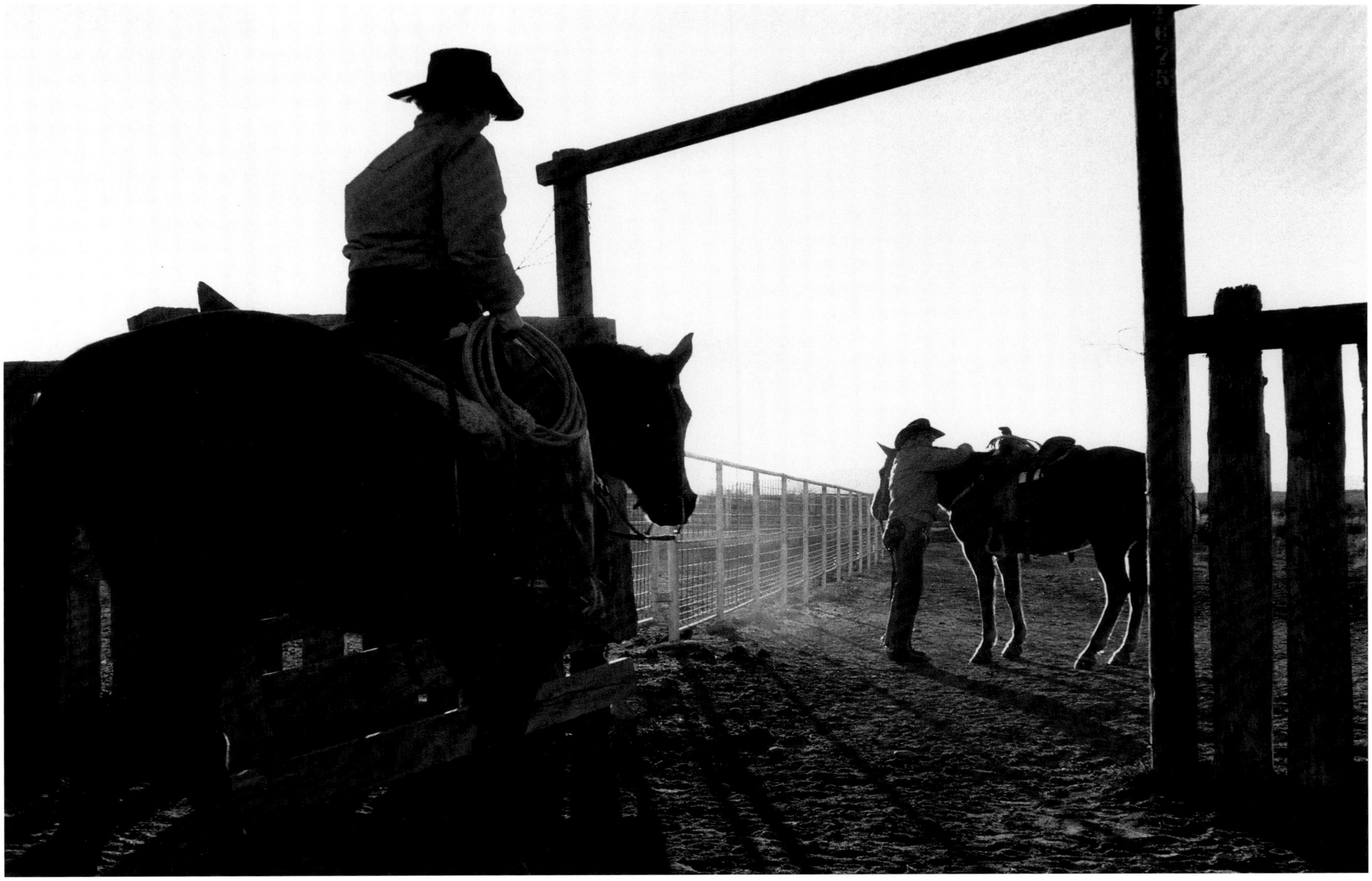

JANE AND BEN CAIN, BAR CROSS RANCH, ENGLE, NEW MEXICO
1987

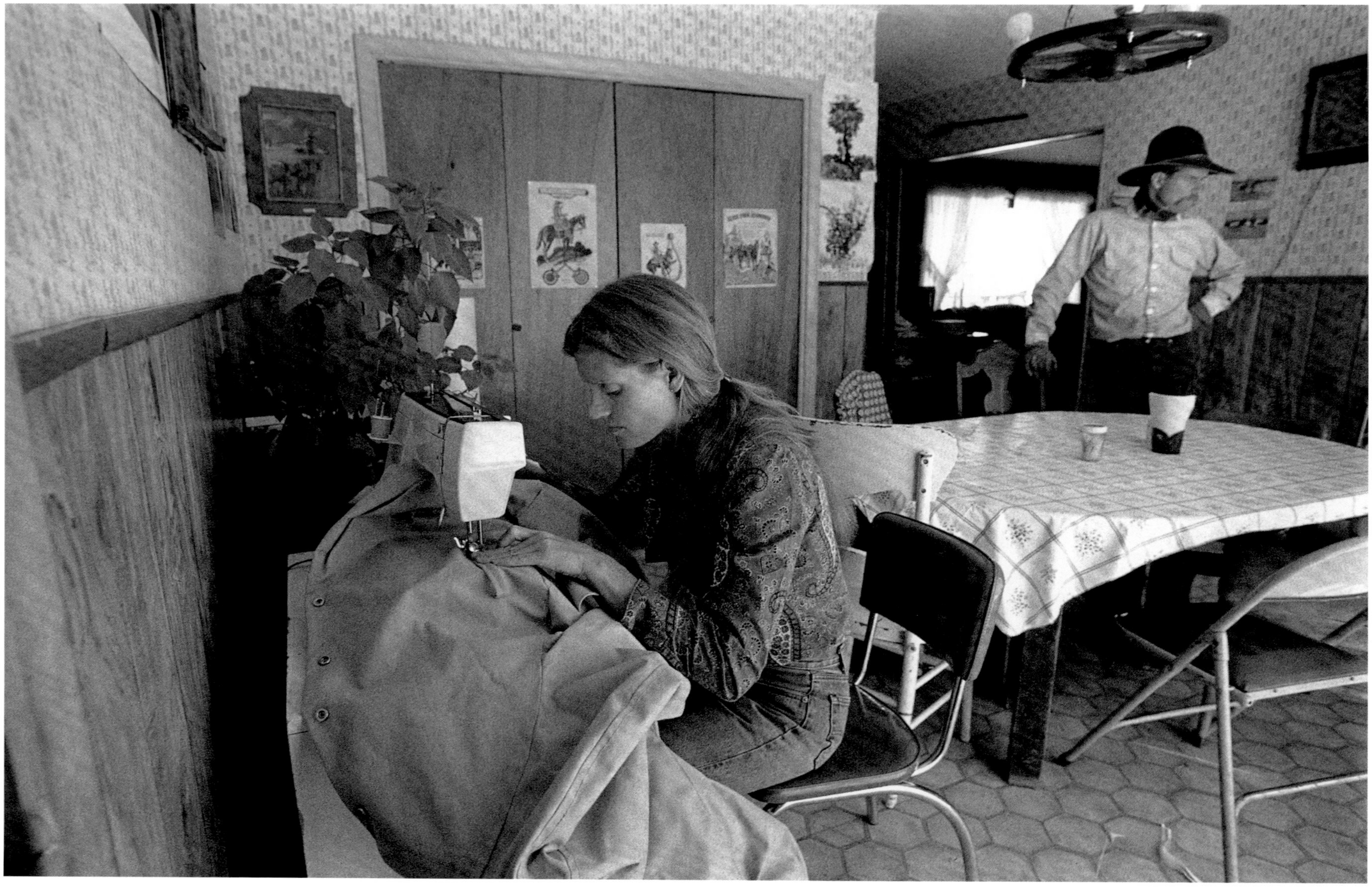

"TOOTIE" MITCHELL HANSEN, SEWING A DUSTER FOR HER HUSBAND,
STAKE RANCH, ELKO, NEVADA
1986

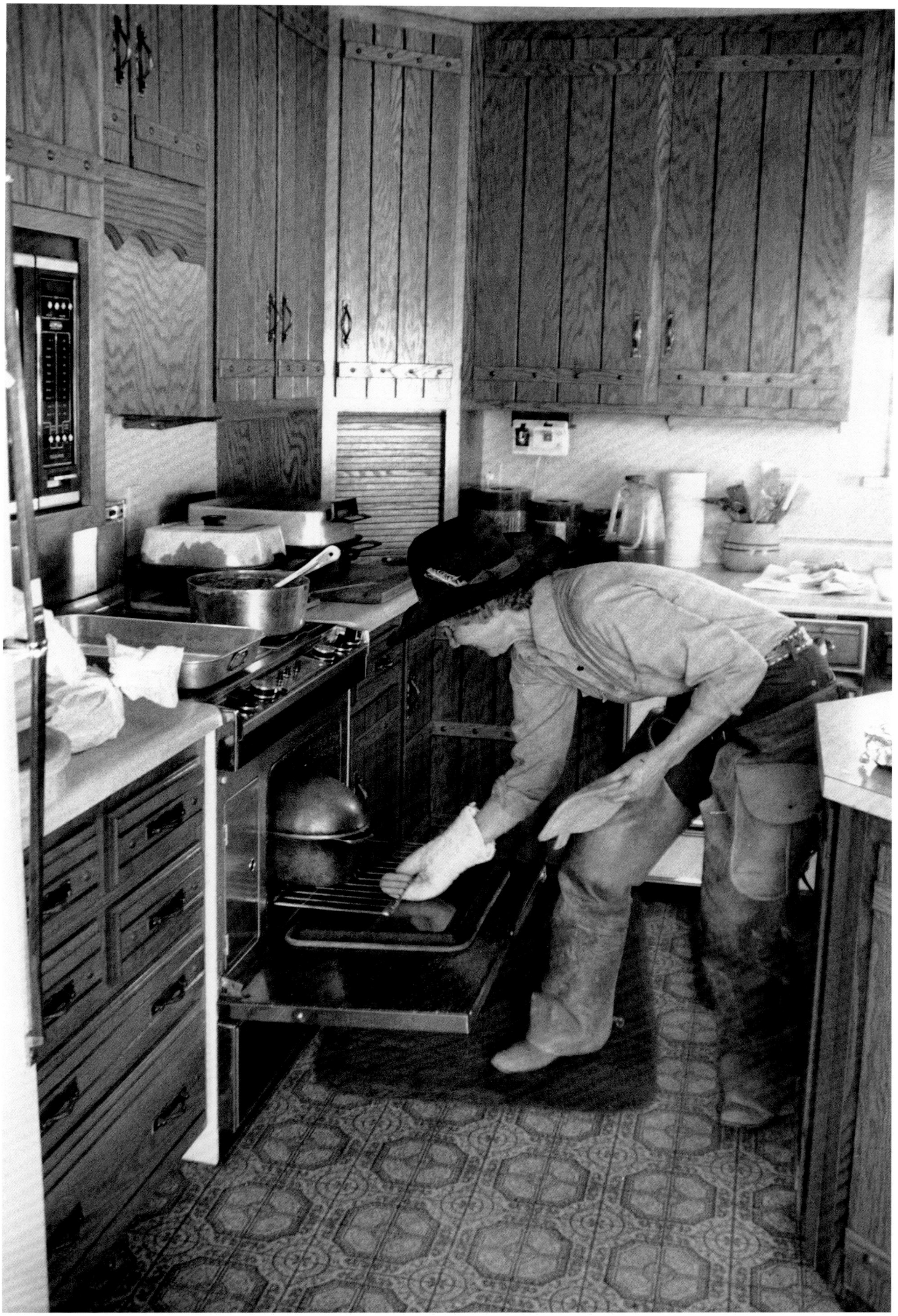

JANE CAIN COOKS FOR NEIGHBORS AND CREW, BAR CROSS RANCH, ENGLE, NEW MEXICO
1987

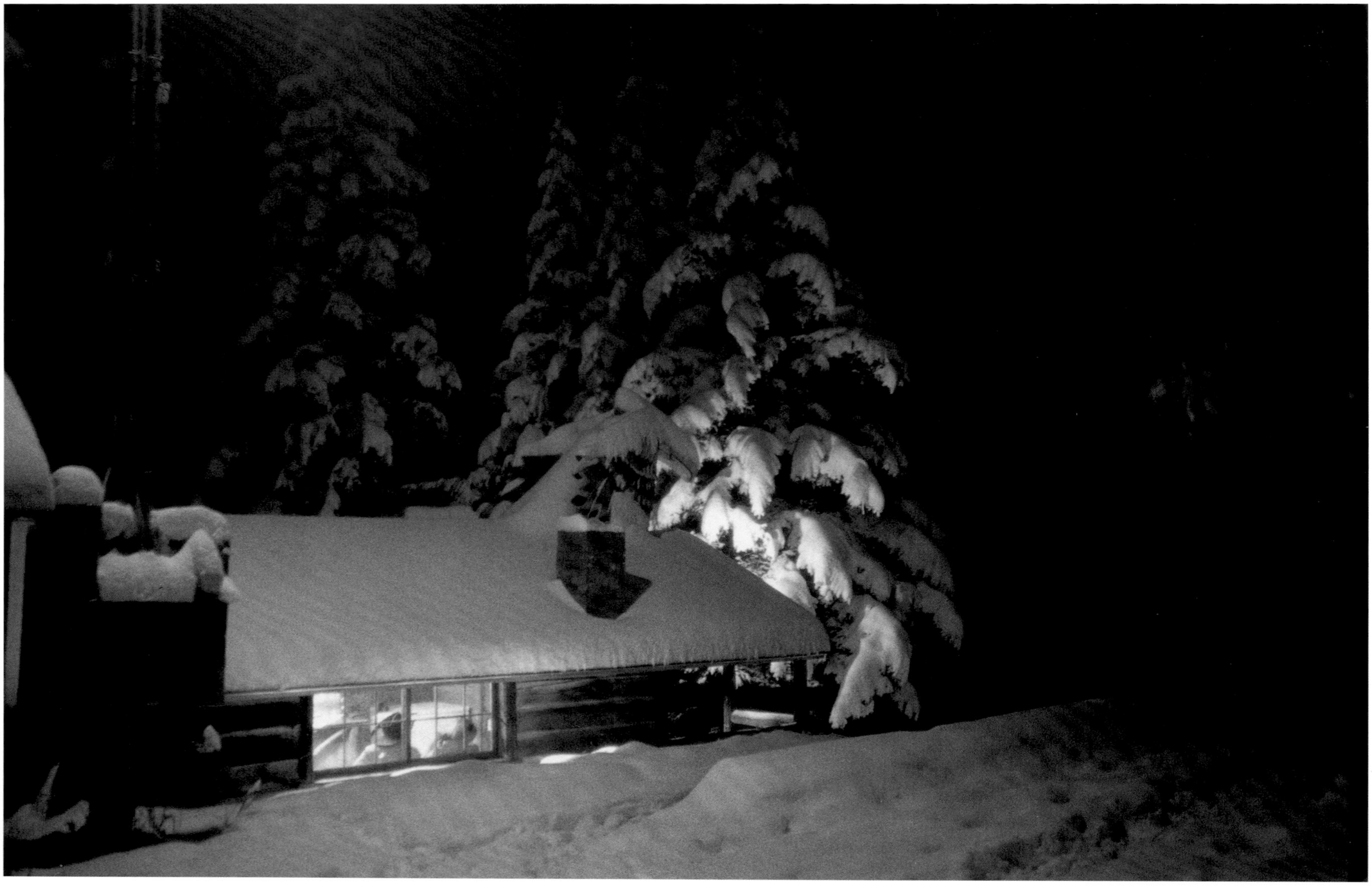

"WHEN NIGHTS ARE SHORT"—EARLY WINTER UP BIG TIMBER CANYON, MONTANA
1982

At the Head of the Creek

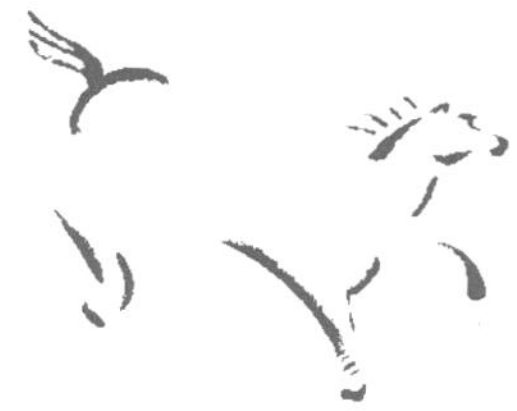

Barbara and I quit college—we were too far apart and I was afraid somebody'd undercut me—and got married in the fall of 1934. It cost me a Harvard degree, but hell, lots of men have those, but I'm the only one who has Barbara. I got 'way the long end of the trade! We spent the next three years in the mountains at the dude ranch by ourselves from September to May. We wintered on thirteen dollars in '34–'35. Was all we had, so we made it do. It bought flour, coffee, sugar, baking powder—and cartridges. It also got us a gallon of moonshine—for medicinal purposes—from a friend. We had anyhow half the jug left in the spring, too. We also had a couple of sacks of last summer's spuds in the root cellar.

We lived well and happy. We might have lived better if I'd known as much about beans as I do now, thanks to the time we spent in Arizona later, but it never occurred to us that we were poor. If somebody had told me so I'd have whipped him, or damn sure tried to. We just knew we had to be careful of what little we had, so we did fine, because in my book "poverty" depends an awful lot on a man's state of mind, not just what he's got. Rifle shells were precious, and it didn't take me long to learn the cardinal rule of a good poacher—don't shoot until you know you can kill what you want to kill, clean. A rifle shot makes an awful lot of noise, too, when it's illegal. We lived on game those first three winters, and Barbara learned to cook meat by cooking venison, which is tricky. The result is that she's the finest meat cook, any meat, I know. She did such a hell of a job with it that we *still* like deer meat; matter of fact I've never seen the time of year when it wasn't good. I got a big old bear the first year, too, and Barbara still claims bear grease is the finest shortening she's ever used. The meat was fine and a welcomed change. It tastes

SPIKE VAN CLEVE (1912–1982) PUBLISHED "AT THE HEAD OF THE CREEK" IN 1977 IN THE COLLECTION *FORTY YEARS' GATHERIN'S.*

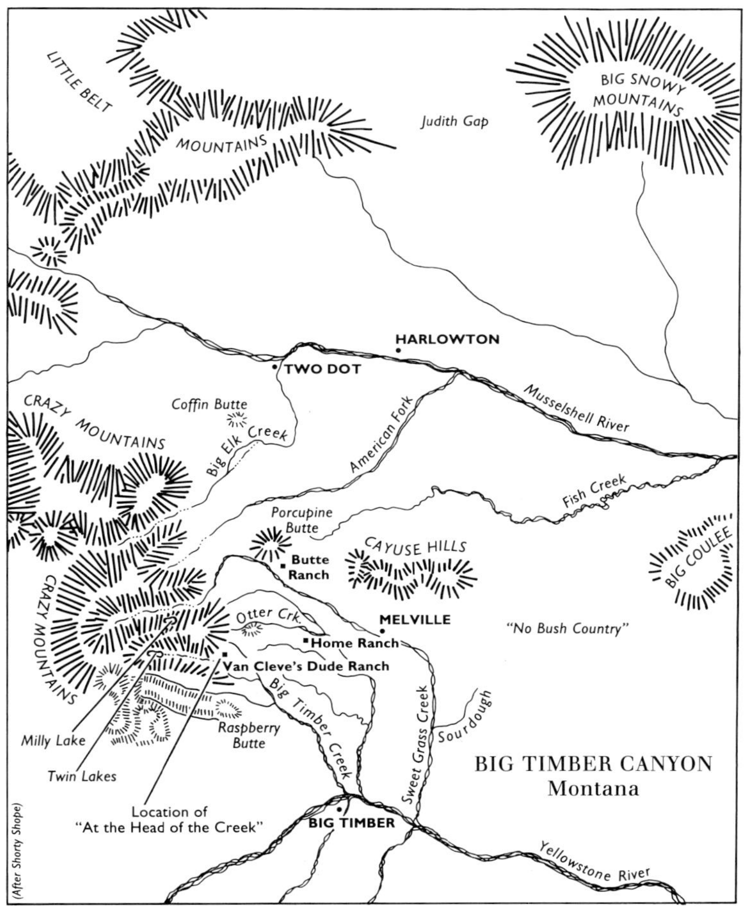

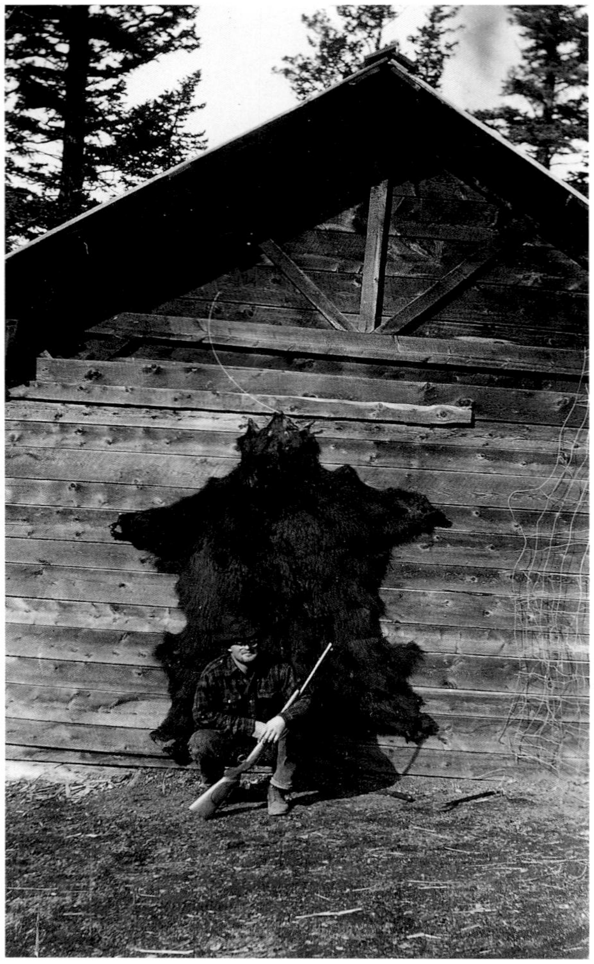

THE SECOND WINTER OF MARRIED LIFE. SPIKE VAN CLEVE WITH HIS 522-POUND BLACK BEAR, OCTOBER 1935.

very much like pork, but is dark. Took her some time at first to get over a remark that an old boy who was working for the Forest made when he saw the skinned animal hung up: "Damn if it don't look jest like a man hangin' there, though them legs an' arms is a shade short." Damned if it didn't, too!

I remember an incident along in March when we were out of meat. I lay out on a windy ridge until I could pick out a dry doe and get a good shot at her with my last rifle shell. I hit her, but the damn wind fouled things—either that or I was so bone deep cold I was shaky—and I didn't hit her where I wanted. I knew that if I followed her right away and warmed her up, she'd quit the country. So I went home, saddled up and rode about six miles to a neighbor who had a phone; ours didn't work in the winter. I called Mother and asked if she'd drive down to Melville and get me a box of 25-35s, and that I'd ride on down to the home ranch and pick them up. She agreed, so I jogged on down another six to eight miles to the lower ranch, picked up my shells, rode back to where the deer was bedded down, slipped up and killed her. When I dressed her I left the heart, liver and other goodies attached, tied her behind the saddle and rode home to the canyon ranch, and we were in business a while longer. To this day I hate to waste shells and I *won't* waste meat!

That first winter about all I got done was getting out wood for our box stove and the cook stove, which were all the heat we had in our two-room cabin, and packing water up from the creek. Big Timber Canyon is narrow, and the buildings are well up the hill from the creek.

I'd skid timber out with Jack and Kelly and cut it up with the only saw on the place, a good Disston two-man crosscut. I figured out how I could hang a weight on the yon end of the saw in place of another man and got by, though it was tough sawing. The first piece of equipment I bought after we were married, and I charged it, was a seven-pound Kelly felling axe, for I had expected to get out some fence posts. However, my two household chores precluded that, but I put the axe to good use anyhow.

The blocks I sawed I split with a maul and wedge. That fit the box stove, but for the cook stove I had to split the blocks again into lighter wood. Then, finally, I'd pack it down to the woodshed I built on the front of the cabin. If I worked pretty steadily at the wood pile I managed to stay a little in the lead.

Of course, since the ranch was a summer outfit and we normally didn't have anyone wintering there, the pipes were either on or near the surface of the ground and had to be drained so they wouldn't freeze. It was quite a climb up from Big Timber Creek with the two buckets I made by cutting the tops out of a couple of those old-timey five gallon gas cans and putting wire bails on them. Ten gallons of water gets to be a load on a snowy sidehill. It was that winter that I told Barbara that woman was man's natural enemy when it came to his ability to split wood and pack water. I found I hadn't seen anything, though, when our daughter Barby arrived the next year!

That second year we were married, '35–'36, was a rank winter. The temperature never got above twenty below zero for almost two months, and one really brisk night it hit sixty-eight below! I saw moon dogs that night and heard trees exploding, even over the creaking and popping of the logs in the cabin as they twisted and shrank from the cold.

That fall I'd had time to plow a furrow from the spring box down to the cabin for the water pipe. I got it in and covered, and for a while we did fine by letting the water run in the sink all night. The brutal weather put a sudden end to that, though, and now I not only had to carry water for us, but for a baby, too, and I soon found that a baby uses more water in a day than a grown man, ten to one. It worked me to a frazzle, for there weren't any disposable diapers in those days. We couldn't have afforded them even if there had been, anyhow. All of us damn sure didn't bathe every night either, for it was a complicated proposition.

First I'd fill a wash boiler on the cook stove. When it was hot enough the two of us would pour it into a wash tub on the floor in front of the heating stove. Then Barbara'd bathe Barby. After that it was her turn at the tub, and then mine. When we were all done she saved the bath water and used it for washing the floor. She's never forgotten either, and right today can go farther on a bucket of water than the average woman can on a barrel full!

A couple of more baby-related problems arose, too. One that bothered Barbara some was that every time we lit the candle to give Barby her two o'clock in the morning feeding some kind of a cat would never fail to squall outside. Either a lion or a lynx, for though the tracks were plain I didn't know the difference. I could never get a shot at whatever it was, and Barbara finally got used to it.

It was the oranges that were the real problem. We'd gotten them so Barby would have some fresh fruit or juice, as the baby book said she must. It was a cinch to keep them cool enough so they'd keep—we put the box just inside the cabin door—but when it got so ungodly cold, keeping them from freezing was tough. There was no coal to bank the cook stove, so I tried doing it with green wood and putting the fruit up in the warming oven. It didn't work, for the stove got cold enough before morning that they frost bit. Finally, in desperation, Barbara and I took the oranges to bed with us, and, by God, they didn't freeze, though I must admit they were sort of lumpy bed partners.

The cold sure was a problem, but we kept the baby warm with a hot water bottle and lots of blankets, though many a morning we found a thick layer of frost from our breathing on the spread or quilt by our heads. The damn spuds we had in the root cellar frost bit, too, and though I picked them over carefully, we ended up eating frosted potatoes. For anyone who hasn't done the same I can state unequivocably that the slight sweet taste that frosting gives them gets to be horrible! Like mountain goat meat—a man gets a lump in his throat as soon as he smells it cooking; a lump that grows and gets more unswallowable by the minute. However, we did manage to sort of disguise things by laboriously cutting the spuds into shoestrings and deep fat frying them in bear grease. With a liberal dosing of ketchup left over from the dude season, we were able to get them down. It was a lot of work for damn poor results, and if we'd had anything else to eat in the way of vegetables we sure wouldn't have bothered with them.

I learned that girls were different than boys pretty early, but I had no idea how much different until we spent those three winters in the mountains. Barbara was not a ranch-raised girl, and the adjustments she had to make to the life we led were tremendous. We often laugh now about how she used to figure on a whole day to bake or churn or wash clothes back then. Now she does them all at once, it seems, and has time to work stock with me, to boot. It was the little, funny things I learned about girls that I've never forgotten—feminine mystique, feminine vapors, or whatever, which I had never run up against before. Take the first cake Barbara baked. It either fell or maybe she took it out of the oven too soon, but we didn't have flour and stuff to waste, so she decided to use it anyhow. So she made a good thick frosting and leveled the top of the cake up with it, figuring, as she admitted years later, that I'd never notice what had happened. I didn't, either! When she cut the cake her skullduggery showed, much to her chagrin, but it made no impression on me—I just thought she'd iced the cake pretty thickly and unevenly. I don't like dry cake. This damn sure wasn't, so I told her, completely truthfully, "Damn but this is fine cake, Honey! Just the way I like it, good and chewy. It'll stick to a man's ribs."

There was no answer, and I suddenly noticed she'd left the room. I had another piece of cake and said something again, and thought I heard a sob. So I hurried into the bedroom and found Barbara lying there weeping.

"What's the trouble, darling? Don't you feel well?" I asked anxiously.

Her reply, between sobs, threw me completely. "You didn't like my cake. Said it was horrible, all chewy and sticky, and I worked so hard to make it for you."

What the hell was a guy to do? I'd told her I liked it, and dammit, I had. I'd been absolutely truthful in complimenting the cake. It sure took some explaining and comforting.

Then there was the time Barbara told me she thought we were eating too much venison and bear meat, because she seemed to get queasy in the stomach when she cooked it. It worried me, so I rode down to a neighboring ranch, borrowed their phone, called up Dad and asked if there was any chance I could get a butcher hog from him. When I explained why, he chuckled. It was a party line and he wasn't the *only* chuckler—the whole damn country knew Barbara was pregnant before either she or I did!

The second year I somehow found time to get out some fence posts. Maybe I had gotten so efficient at getting wood and water that I was saving time. Anyhow, I cut timber into six-foot lengths, split it into post size, and sharpened one end of each length with an axe. However, at three cents a post, which was all a man could get for them then, I decided the pay sure as hell didn't match the work, so I tried trapping.

There were no beaver as far up the creeks as we lived, for they need cottonwood or willows for food and we lived in evergreen country. Besides, they were illegal, though I don't honestly think that would have saved them from my depredations if they had been handy. Weasel were hardly worth skinning, mink moved out of the mountains in the fall, and I had learned a long time ago that I'd starve to death in a hurry if I tried to make a living trapping coyotes. But I knew there were marten in the mountains, for I had seen them once in a while up around the lakes during the summers.

Marten are a North American cousin of the Russian sable, and like most of the weasel family are pretty easy to trap, which gave me a fighting chance. Getting to where they ranged was a horse of a plumb different color. The snow ran from three feet deep in the timber at the ranch to right at twenty feet up at the lakes—at least I blazed a tree at snow level up there in February and the blaze was anyhow twenty feet above ground when I located it the next summer—but I figured out a way to handle things.

I'd ride my saddle horse just as far as he could buck the snow, then picket him and give him the sack of hay I'd carried on my saddle. From there on I used an old pair of snowshoes that had been around the ranch for years, but it was no cinch. I have always been a great believer in my granddad's contention that "if God had wanted man to walk He'd have given him four feet, but He gave him two—one to put on each side of a horse." On top of that, the webs were those wide, long-tailed Maine propositions that made a man walk all spraddled out, which isn't the ticket for a steep country. The combination stove me up something fierce at first. I stayed with it, though, and finally got so I could make the round trip in a day. Not easily, but I *could* make it—if I rested up for a couple of days afterwards, that is. So I eventually fell into a pattern of making the rounds of my traps every three or four days. Between trips the wood and water kept me out of mischief.

It was about five miles each way, at least three of which I had to make afoot, and during those three miles I gained some two thousand feet in altitude. So I traveled damn light. My traps I had scattered by horse before the snow got bad, so all I carried was a hand axe and a bottle of scent. The upper country widened out a little, but for a mile and a half the gut of the canyon was narrow, high on each side and steep, so snowslides were a threat. As a matter of fact, after a storm or the wind had built snow cornices on the sides of the cliffs and along the tops of the canyon rims, it was damn dangerous. Any vibration could pop them loose, and I'd sure have been a gone gosling if I'd fired a shot down below them. I didn't carry any kind of a gun—there wasn't anything back in that snow country to shoot at anyway. In places I was even leery of snapping a branch, coughing, or using the hand axe. It wasn't as bad when the cornices were solid, so I always got away early and made the trip up past them before they had much chance to thaw and loosen.

The trips weren't cold really, for the climb kept me warm, but they were pretty lonesome. I sort of enjoyed them at that. The high country, which I had never seen before in winter, was absolutely lovely. On a clear day the peaks lifted up and up, regal, splendid, and though not a whisper of sound reached the canyon floor, they often flew plumes of snow as the upper winds lashed their heads. Sometimes I would stop and just look at them—pure white except where they were too steep for the snow to cling. Mostly, though, they'd be looming dimly in the frost haze or peering through the curtains of a snow squall. I got caught in heavy snowstorms—times when the quiet would be deafening and the alpine firs so completely stock still under their white burdens that I got the feeling that they never, ever moved and that the flakes drifting straight down and I were the only moving things in the world. But then I'd begin to wonder what the hell I'd see if I looked back along my track. I never did; something might be there, and I'd rather not know what it was.

Once in a while, toward the middle of the day and after I'd reached where the upper country widened a little, I'd hear a snowslide in the canyon below me. There'd be a crack like a far-off rifle shot, followed by a sullen mutter that stirred the air, and a distant silvery cloud would blossom above the timber on the flank of a peak and race down into the canyon. Only once was one ever near me, thank God, for I damn sure didn't enjoy it. I was just below lower Twin Lake when the thing let loose high up on my side of the canyon and maybe two or three hundred yards ahead with an explosion like a charge of powder. There was a blast of air and a roar that shook me so that I had to grab the small tree beside me, and a choking smother of snow-filled air churned over me. When it had settled and the noise had stopped, I tiptoed—or as close to it as a man can on webs—gingerly forward and inside fifty yards reached the edge of the slide.

It was nasty! A swath at least a hundred yards wide had been cut through the timber and was littered with broken trees and torn up boulders. The tongue crossed the creek and had driven a couple of hundred feet up the opposite slope and ended in a welter of trees—thrown at all angles like jackstraws—boulders, and snow, from which came unpleasant noises as everything settled. As I looked at it I wondered just what in the hell I was doing back there anyway, and pretty and pelts suddenly weren't so all-fired important. I headed for home.

It was earlier than usual when I reached the head of the bad part of the canyon. In my considered opinion too early, by God, to tackle it, so I built a little fire and hunkered over it until I figured things had stiffened up. I was pretty gun-shy about slides right then, and have been a little ever since. Even so, though mountains can be damn dangerous, to me, who grew up at their feet, they can also be friendly and strangely protective, but only if a man respects them. They demand respect. Get careless and they are unforgiving. I have packed out what was left of a careless climber, and I've seen a man who was finally dug out of a slide. There are two things that damn sure have old Spike's respect—mountains and horses.

I learned a little more about girls from one of those trap-line safaris. My saddle horse, Irish, liked to buck, and when he did he'd shed me. I found I could cheat him in the deep snow, though, so I used him. After about one or two trips he slacked off, and even though I got into the habit of putting more odds and ends on him than Carter has pills, he never bobbled. Though by spring I thought he was broke, when the snow went off I discovered he sure as hell wasn't, so when we moved out of the mountains I traded him to Cremer for Sailor, an honest old cow horse. Irish ended up in Cremer's rodeo string, and I saw him

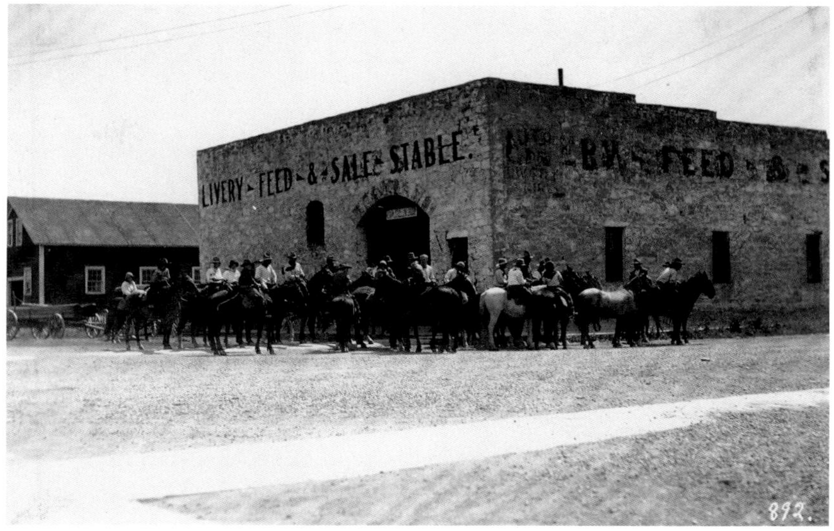

EXCITEMENT ON MAIN STREET IN BIG TIMBER JUST BEFORE THE RODEO, CA. 1925.

put a couple of good riders down at the Livingston Roundup. Made me feel a lot better.

I took him this day as usual, and a while after I got into my webs it began to snow. It got heavier and thicker as the day went on. I had trouble locating my traps and finally it got so nasty it was all I could do to find my way back down the canyon. I made it all right, but it took some time; it was full dark when I rode into the ranch.

A bar of light from the Coleman lamp was trying to make some impression on the curtain of heavy flakes outside the cabin window, and I could see my wife, nose against the glass and hands cupped around her eyes, peering out as we came in. Both my horse and I were so covered with snow that I knew she couldn't see us, so to let her know we were back, I rode up outside the window, waved and grinned. She saw us materialize out of the storm, relief replaced the worried look, and she turned away. I rode on down to the barn, tended Irish, milked our cow, and bedded everything for the night.

When I got to the cabin, I swept myself off in the woodshed and stepped inside. Barbara wasn't in evidence. Supper was still on the stove, and the table wasn't set, but I called cheerily, "I'm home, Mommy, and 'tain't fitten out."

Not a sound, so I walked into the bedroom, thinking she was tending to the baby, but she already was in bed.

"What the hell, Honey?" I asked.

The response stunned me, for Barbara reared up, her eyes blazing, and in a voice about evenly divided between rage and tears poured out, "I've been worrying about you all day long in this storm. You didn't come back before dark like you promised you'd always do. I thought I could hear snowslides. I went down to the barn to see if that damn bronc you insist on riding had come in by himself. I rang the bell every few minutes, but the heavy snow muffled it. I didn't know what I'd do anyway if I found your horse. We haven't a phone. You let *me* worry all day and *you* knew you were perfectly all right all the time, damn you. I've had supper ready for hours and I don't care whether you eat or not. Just go away!"

My abject apologies for being perfectly healthy and a manful attempt at sweet talk were miserable failures, and I ended by eating my dried-up supper alone. I washed the dishes and, hoping for Brownie points, made plenty of noise as I did so. Then I sneaked into bed to be thrust firmly to the very edge of my own side of the mattress and told bitterly, "Get away. I hate you."

It wasn't a pleasant night, nor was the next day much better, and it was some little time until I was restored to grace. It taught me once and for always, though, that if a girl worries real badly about someone and everything turns out fine, then she'll get on the fight. Be damned if I can understand why, but it's sure a fact, and a smart man learns to live with it—and keep his mouth shut.

It showed up another time, too, when a couple of us left camp on the head of Sweet Grass to do some fishing in Milly Lake. It was a booger to get into, even after we had left our horses, and when dark sneaked up on us to end our fishing we decided it was too dangerous to try to go out until daylight so we lay out all night. We started early the next morning, got to our horses, dropped down the fork to the main canyon and started up it for camp. They had been worried about us up there, so Barbara and Dad left camp to look for us at about the same time we left the lake and very thoughtfully took us each a couple of hotcake sandwiches—they had run out of bread. Well, they spotted us as they came around a bluff, and as Barbara heard us talking and laughing she realized that we weren't in any trouble, so she got the sandwiches out of her saddle pocket and threw them into the brush down the hill.

Dad asked her what she was doing. She told him, and he ordered, "You get off that horse and go get those sandwiches. Those boys haven't eaten since yesterday morning, and if it rained on us in our tents last night, it damn sure rained on them, too. Hike now." And she did! I never had that much nerve, though I suppose a father-in-law can get away with things a husband can't.

While I'm on the subject of "feminine vapors"—a magnificent Victorian description, that—I must mention something else I learned about girls years later. If Barbara ever really painted for war about something, I damn sure better not mention that something again after she cooled down, especially if she happened to have the tiniest bit of legitimate reason. Even years later, she gets as waspy about it as she originally was. Maybe even worse, so I sure never mention any of those—to use a mild word—controversial issues from the past. And I do my best to head off any conversation that is taking a dangerous slant. I believe in letting sleeping, by God, dogs lie!

Funny, Barbara claims she never would have made a pioneer woman, and I say that she *was* a pioneer woman. A town-raised girl plunked down back in the hills a couple of weeks after her marriage, living conditions almost completely primitive, and alone for months on end except for me; and hell, we didn't really know each other when we first were married. I was gone most of every day, to boot. I don't know how she stood those first years in the mountains, and she admits now that she doesn't either. Because they broke the mold when they made Barbara, she not only stood it, but adapted to it and took care of me and the children as well. (Tack had arrived by the third year, a little over thirteen months after Barby.)

When winter really socked in, usually right around the first of the year, the deer, coyotes and almost all the other animals and birds pulled out for lower country, and we were alone except for a couple of saddle horses, a cow, an old three-legged cat that had wandered in and taken up a claim, some cocky little chickadees—who brightened things considerably with their cheerful chatter—and an occasional camp robber. We were set down in such a deep canyon, too, that from November to mid-March the sun never topped the south ridge until at least eleven in the morning and went down behind it again between one and two in the afternoon. Not only that, but clear days were rare, for there was almost always something going on up at the head of the canyon. A squall would build up, work down past us a couple of miles, and then back up, churn around awhile, and then here it would come again. Sometimes half a dozen a day. Or maybe one would just squat down over the place and storm, while a few miles below things would be bright and sunny. Possibly we had one day out of ten that it didn't snow sometime between morning and night, but I doubt it. Then there was the wind; when it came down from the tops into the canyon it was like living in the throat of a bass horn! Maybe I chuckle about some of Barbara's reactions and tease her about things, but it is damn sure based on deep, fervent admiration. Courageous and indomitable, pale words to describe my wife!

Only once did Barbara come apart at the seams. I was off up the canyon and when I got back that evening she was in a state. Between outbursts of tears and fits of uncontrollable laughter on her part I managed to get the story. Then and there I decided we better get down where there were other people.

What had happened was that after I had been gone for several hours and the children were asleep, Barbara was churning when suddenly there came a knocking from the front door of the cabin. Surprised, for we were snowed in and it had been over three months since we had seen anyone, she hurried to the door and eagerly opened it, pleased at the prospect of a visitor. No one was there.

Puzzled, for she was sure she had heard a knock, Barbara went back to her churn. In a few minutes the sound came again, and once more nobody was at the door. The months of solitude began to take over, to affect her thinking. Was somebody trying to play tricks, knocking and then hiding, or was she really hearing anything? Could it be the cabin fever she had heard of, or was she losing her buttons? So she stood beside the door and when the erratic tapping sounded, jerked it open in hopes of catching the culprit. Again a blank, so she ran out the woodshed door and scouted around the cabin. Nothing. Not even tracks, except those I'd made that morning down to the barn and up to the springhouse. Maybe, though, before she had come out of the woodshed, whoever it was had been able to run along my tracks and get out of sight into one of the two buildings. But it was too far to have made it to the barn before she appeared. The springhouse was closer. Perhaps . . .

Up to the building she went with no thought of overshoes, coat, mitts, or anything except to find out who was doing the knocking. Once again, nobody, not a thing, but suddenly the knocking came again from the woodshed on the front of our cabin.

"My God, my babies are down there," came the chilling thought, and down the path she went on high. Just as she reached the shed door out flew a camp robber, and everything fell into place—it had been the bird pecking at the frozen meat we'd left on the shed floor for our old cat, Peg, that she had heard.

Her reaction wasn't plumb under control by the time I got home, and it bothered the hell out of me. This was no place for a girl and two babies. Today, if my children planned on living like we did, where we did, I'd break their damn necks! But we were young, and probably the finest, most magnificent thing about youth—at least in our time—is the faith that, hell, everything will turn out all right. It nearly always does, too. But a man learns not to crowd it.

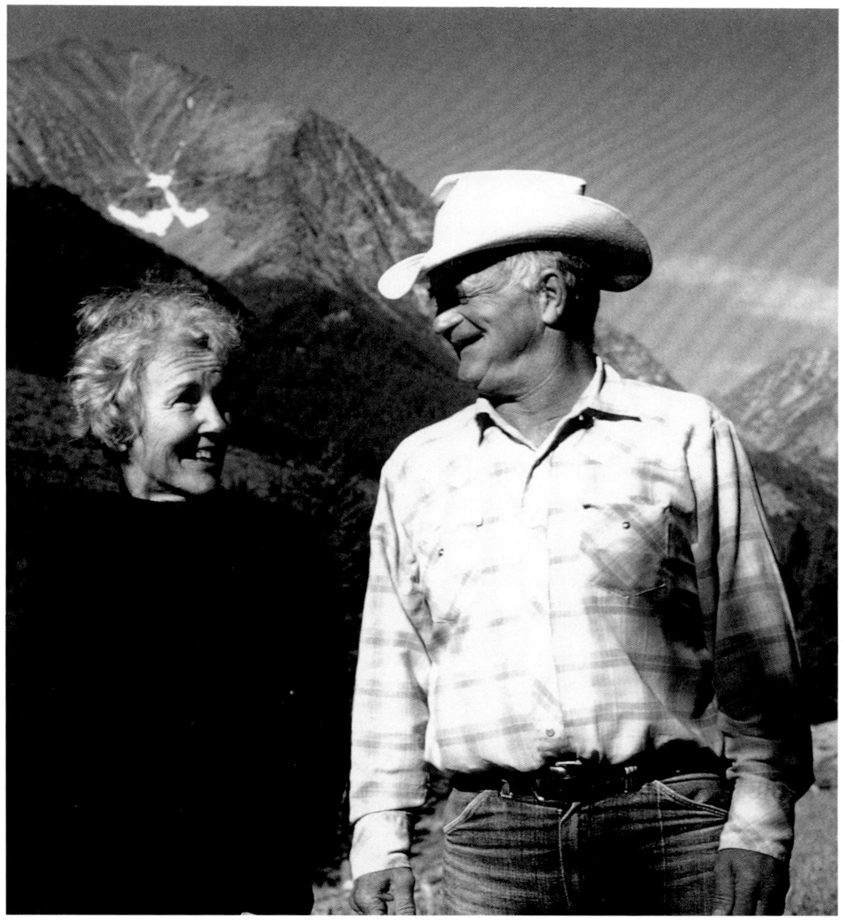

BARBARA AND SPIKE VAN CLEVE, CA. 1975.

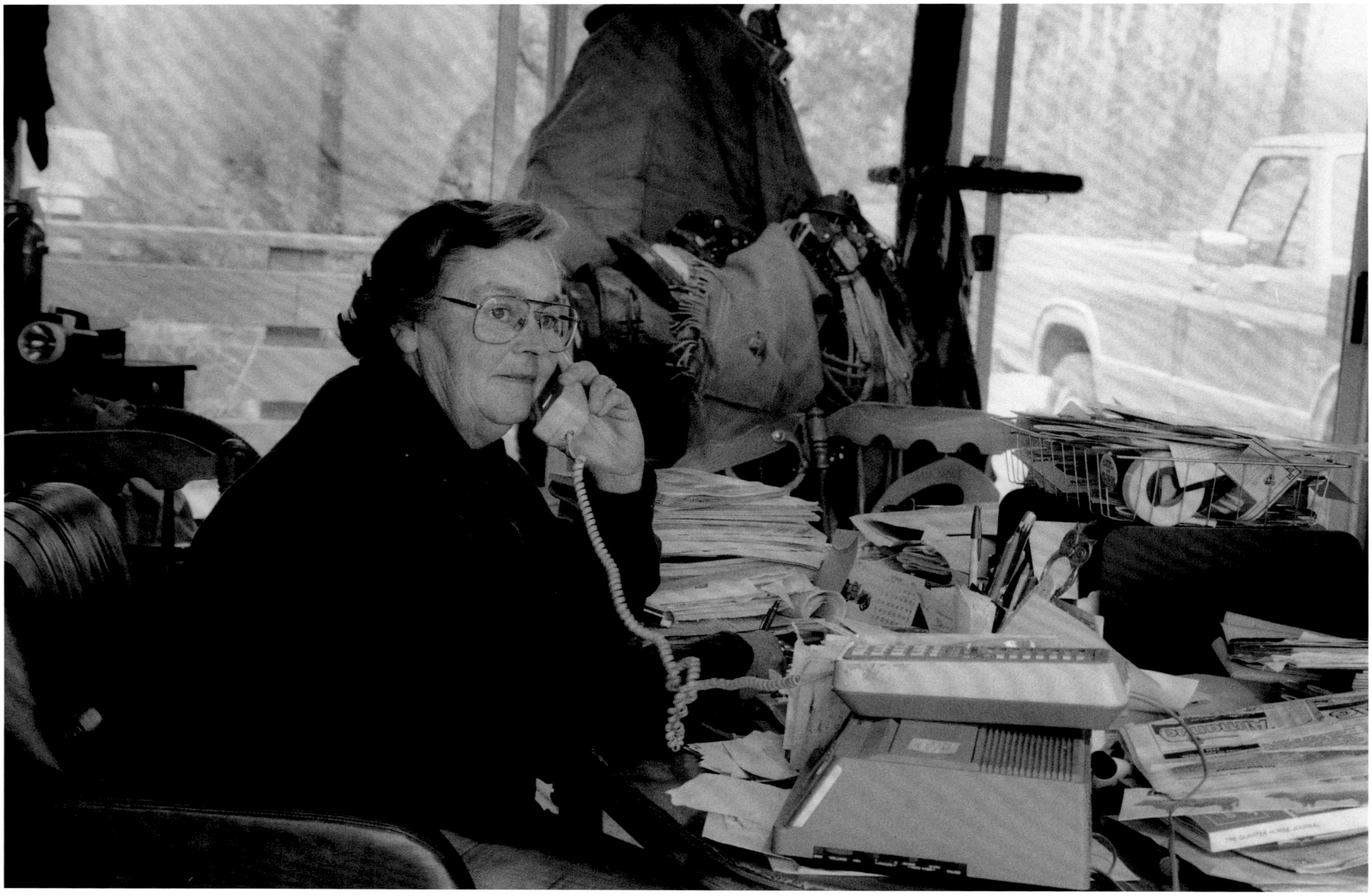

JANE GLENNIE AT THE GLENNIE RANCHES, MARTINSDALE, MONTANA
1987

Western Ranch Women

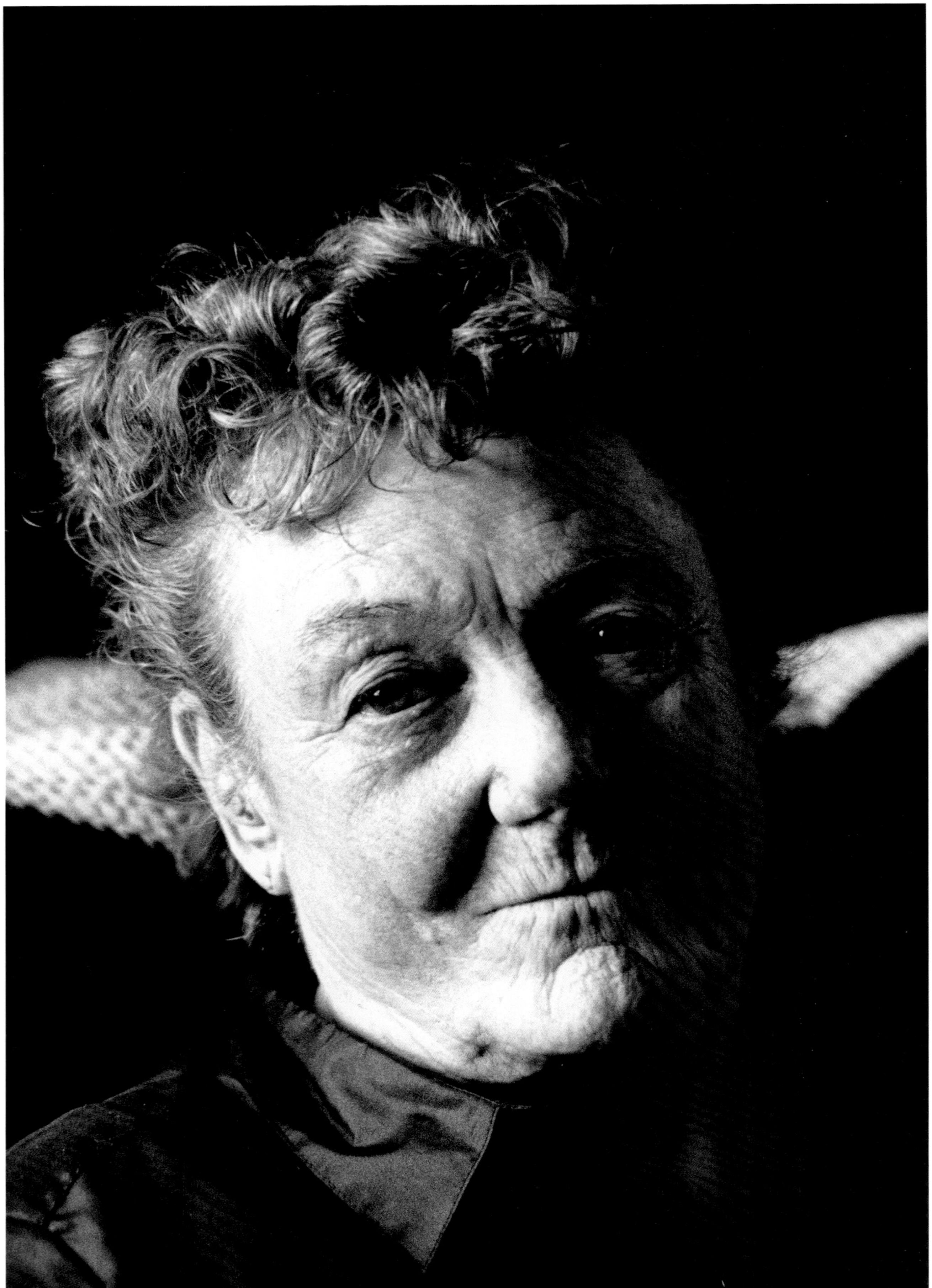

BOBBY BROOKS KRAMER, KRAMER HORSE RANCH
1986

Bobby Brooks Kramer

Kramer Horse Ranch, Billings, Montana

Bobby and her husband, Bud, were both "born fifty years too late." She is about eighty-one now and still trains cutting horses daily, riding them herself. Stove up with arthritis in one knee, it is a labor of love for Bobby. Back in 1989 she rode every day, all day for a solid week on the Great Montana Centennial Cattle Drive, starting well before dawn and quitting after sunset. She was born over in eastern Montana, a great sprawling and wild chunk of country. The days of the open range were well in the past, but most of the country she called home was still unfenced: "I can remember when we would ride from Cohagen to Miles City and never have to open a gate," she says, a distance of about sixty miles. But the days of riding the big circle are over. Progress, I guess.

Bobby is a remarkable woman: one of the first women bronc riders and a first-rate horse hand. She took up flying at fifty-four, is adept at handling eighteen-wheel livestock semitrailers and was owner, with her husband, of the largest horse ranch in the United States—10,000 head! And today she's not sitting back taking it easy. She is more comfortable and at home on a horse than on the ground. Raised on horseback, she never felt inferior to a man:

When I was on horseback, I felt equal to the strongest man there was. I first rode a bronc back in '37 or '38. We were all going to a rodeo at Wolf Point to ride broncs. I went along and lost all of our money in a gambling game at the carnival. To buy us some hamburgers and gas to get home, I rode an exhibition bronc for $25. That was a lot of money then!

Bobby married Bud in 1943 while he was on leave from his cavalry regiment. After World War II, the plains were still thick with horses and "people were wanting to get rid of those horses," she said. Ranchers would sell them to the Kramers for "range delivery," that is, the Kramers had to gather them on the range. They processed four to five thousand horses a year. Each horse netted the rancher between $7.50 and $15. Some of the horses would be sold as "canners" (dog food, et cetera); some would be sold as bucking horses, especially at the famous Miles City, Montana, bucking horse sale; and some mares and stallions would be kept to use as breeding stock for first-class bucking horses. It was a wild and wooly proposition gathering those wild horses.

Most of the time we had between four and eight of us to gather and trail 300 to 900 horses to the Ingomar [Montana] stockyards. It was about a two-day trip. We didn't travel at night but had a set of big corrals half way there where we'd keep them for the night. We tried to make it easy on ourselves if we could.

I really do miss gathering those wild horses, but it is a young person's game. There was a lot of hard riding and you had to be conditioned to handle it. We'd travel at a long trot because at a long trot a horse can go further and cover more country and still be able to do all the work we had to do.

It had a romance to it that you can't describe. I loved the wild, free ways of the horses. Running horses is one of the greatest feelings, a free, kind of wild feeling! It is almost indescribable. It is like a Chinook wind [warm wind in the winter that melts the snow suddenly] that blows across the range. There is nothing like that feeling or smell. And now it's all gone. It'll never be back.

Bud died in 1979, after which Bobby and her son, Gary Crowder, established a horse ranch outside Billings, Montana, where they raise quarter horses and train cutting horses, with buyers coming from around the world.

The joy of being able to enjoy a good saddle horse, to utilize him working cattle and to do what you need to do on horseback, is one of the greatest feelings in the world.

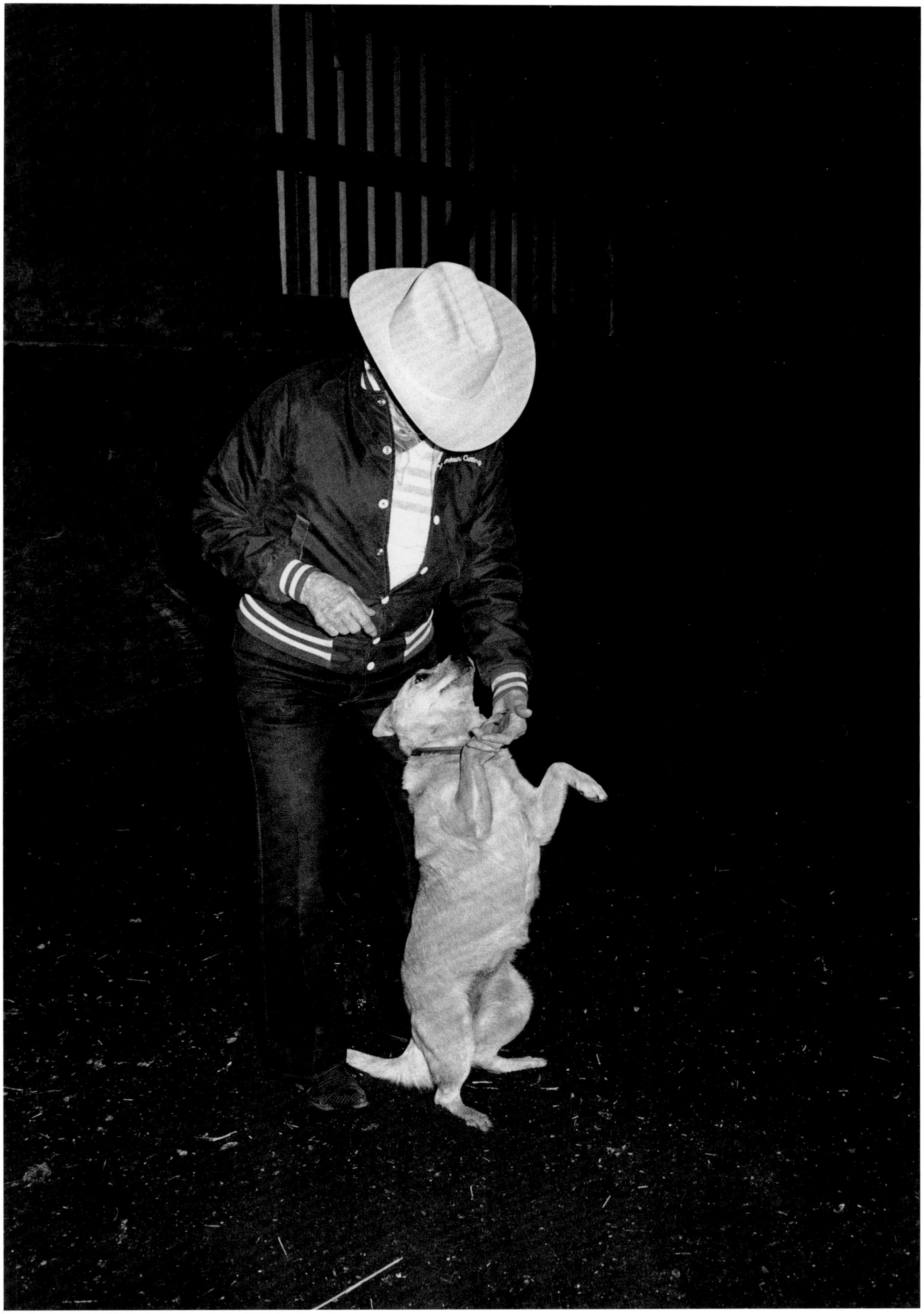

BOBBY BROOKS KRAMER, KRAMER HORSE RANCH
1986

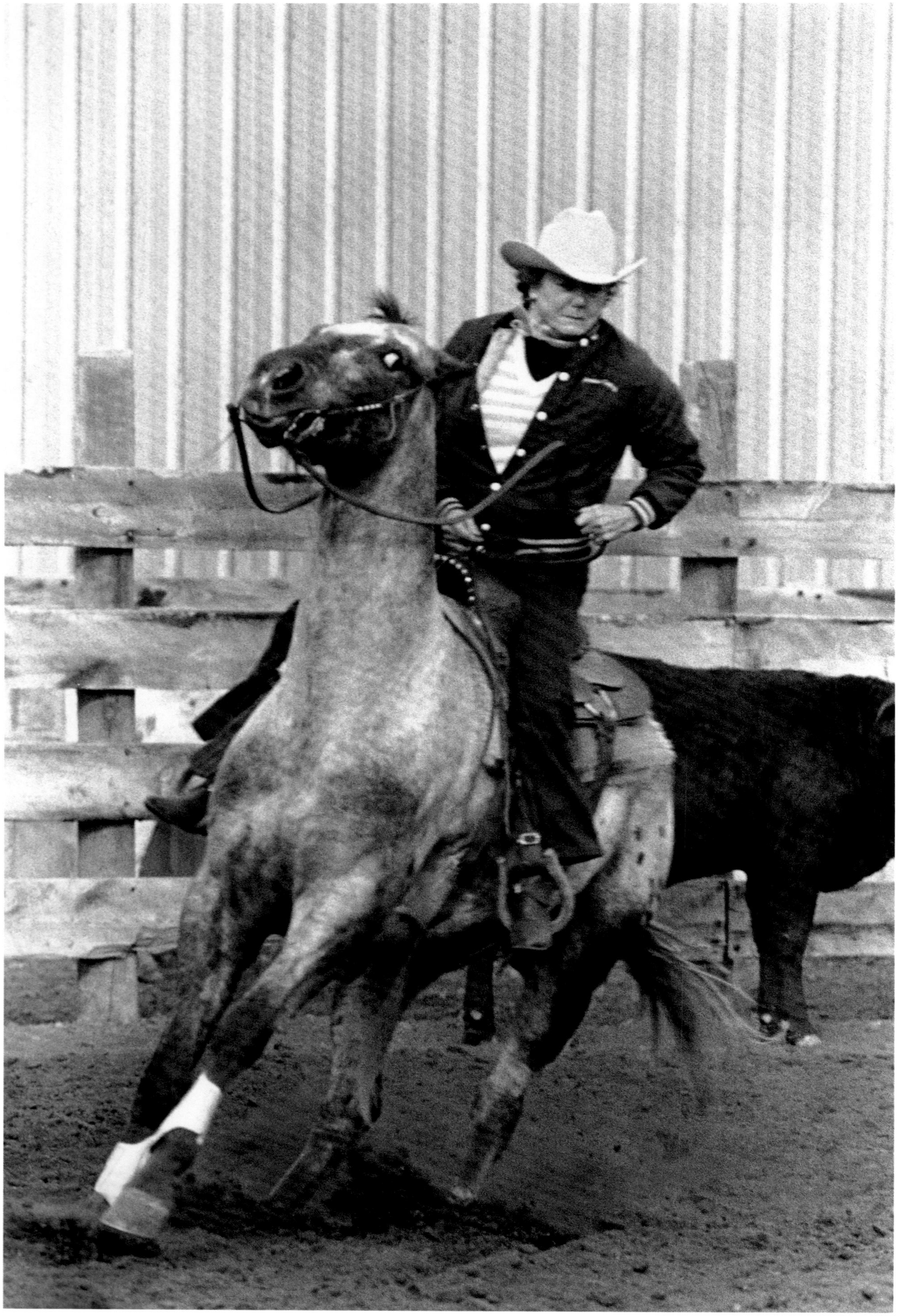

BOBBY BROOKS KRAMER, TRAINING A YOUNG HORSE, KRAMER HORSE RANCH
1986

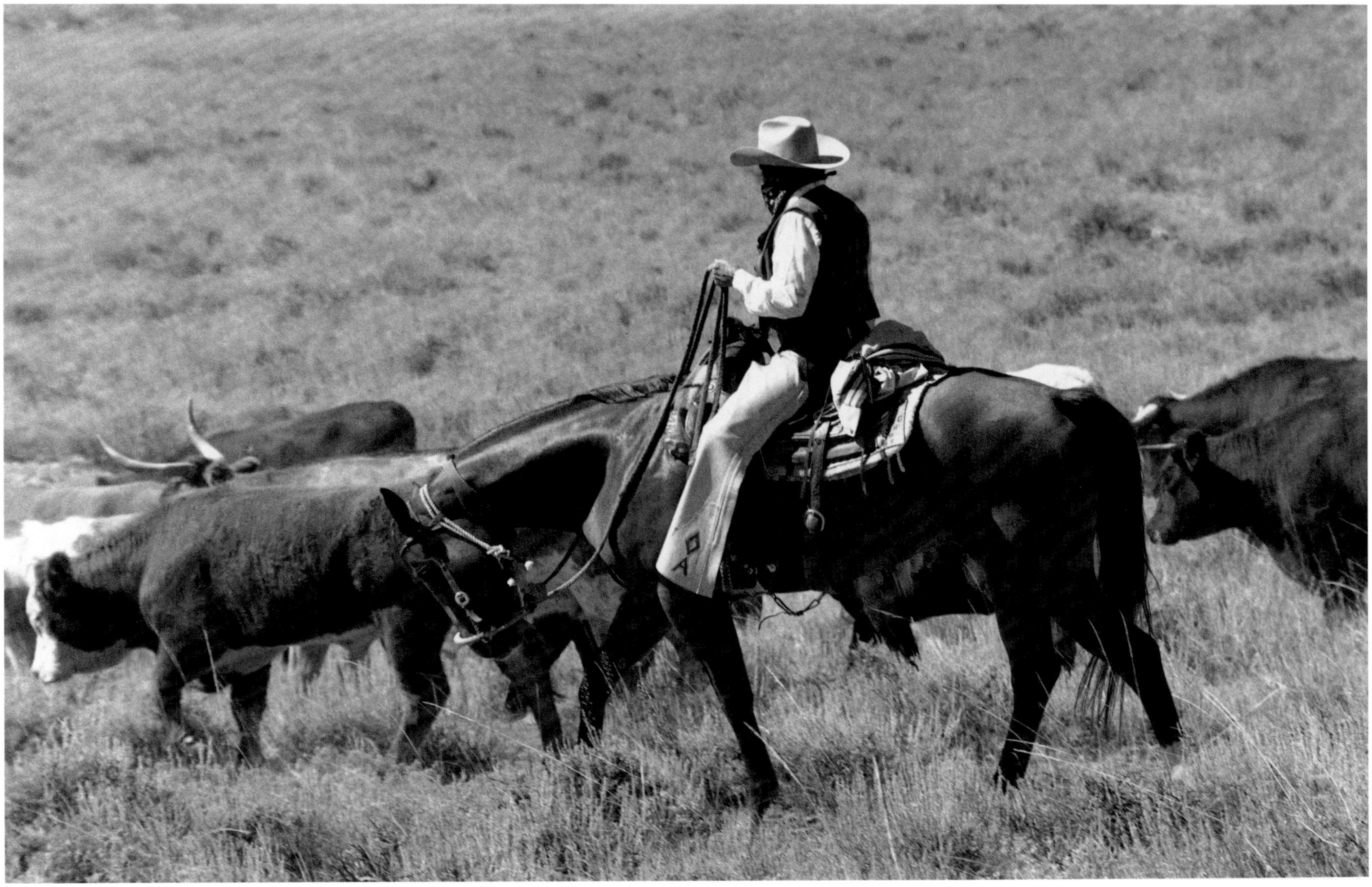

BOBBY BROOKS KRAMER, KRAMER HORSE RANCH
1986

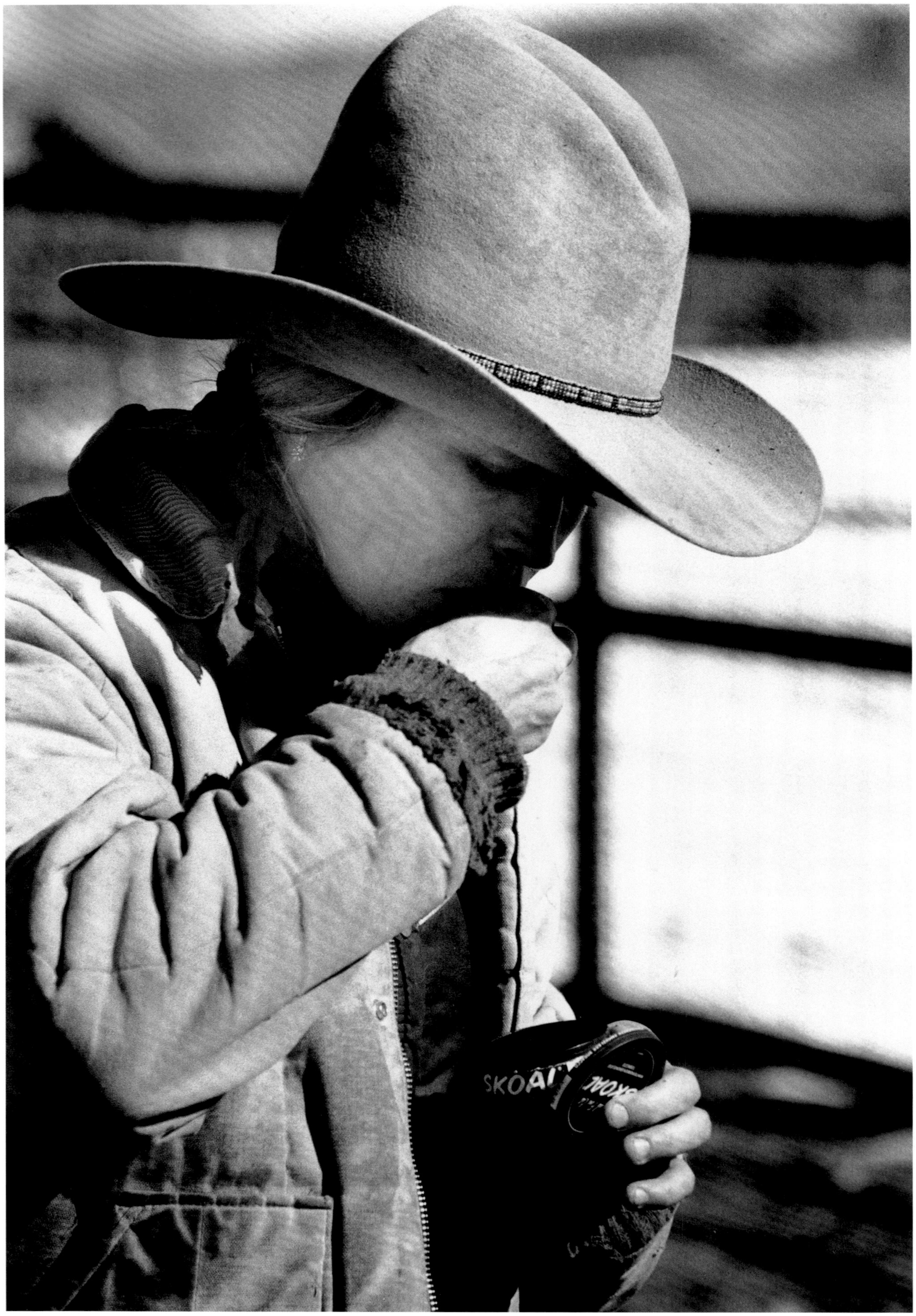

MELODY HARDING, BAR CROSS RANCH
1986

MELODY HARDING

Bar Cross Ranch, Big Piney, Wyoming

A petite, pretty woman of thirty-five, Melody Harding grew up ranching with her family. "I was an only child, so to fit in with the scheme of things I just worked. I didn't want to be left out in the cold so I participated. That was just the way things were done." Finally, she decided that it was the only life for her and began to hire out, working as a ranch hand in the eventual hope of having a ranch of her own. She is so good a hand, so careful in the handling of the ranch owners' equipment and machinery, so knowledgeable about cattle and horses, the land and kind of grasses each range produces, that she is a "cowboss," or ranch foreman, and is much sought after.

Although small in stature, Melody is a "hard-twist," five-foot, four-inch mass of muscle and sinew, like rawhide that is dried or cured. Gentle with all animals and feminine in the best sense of the word, she shrugs "so you age a little faster [working on a ranch] than you would normally—that's the least you can pay for a good life. At least you're in a wholesome atmosphere and you're doing something worthwhile." She always wears earrings and uses chewing tobacco, or "snoose" as it's called by most ranchers.

Melody had been married for twelve years because, as she tells it, "he kind of asked me when I was preoccupied doing something else on the ranch and I said 'Yes' without even thinking." He fell in love, she didn't. They're now divorced. Shifting thought gears she says quietly, "My animals are probably one of the closest things to me; the horses and the dogs are with me daily. It's going to be tough when I lose any of them. Last winter, my oldest dog really lost a lot of ground. It was depressing to watch his frustration because he couldn't go get the cow farthest out and run her back into the herd."

West and a little south of the Wind River Indian Reservation lies the little town of Big Piney, which is situated at the north end of the Fontenelle Reservoir. Much of the land south of town and along the area of the reservoir is subirrigated and has a tendency to develop bog holes and "bottomless" areas when the snow melts or there are heavy rains. It is rolling hill country with mountains to the north.

I visited Bar Cross Ranch during a typical March. Earlier there had been a spring thaw and, as a result, some bottomland had become soggy. The hired man had tried to cross it with the truck to feed cake and hay to the cows and their calves, but he had buried the big four-wheel-drive truck up to the transmission case, even with chains on all four tires. Unable to get unstuck he left it. A snap freeze hit and that truck was frozen in, solidly. Melody brought out the huge Massey Ferguson tractor to pull it out, but after a couple of tries and seeing the leaf springs of the truck bend to what we thought was the breaking point, she decided to wait a few days for a thaw. After a good "Chinook" (warm, thawing wind), she was able to pull it out. In the meantime, she fed by hand with her favorite team, Miff and Muff, a time-consuming alternative to feeding with the tractor.

Melody usually feeds the cattle five rolls of hay a day, each roll weighing 1500 to 1800 pounds. Because of the size of the bales she uses a big tractor with a hydraulic lift spear to carry each bale and an auger that cuts up the roll for easier consumption by the cattle. As we haul the last roll out she muses,

Once in a while you need plain brute strength, so being a woman can be a disadvantage, but that can be turned to an advantage if you can figure out a way to do the job without using brute strength. And sometimes you come up with a really neat alternative and usually don't tear up as much stuff as you would with the brute strength technique.

She also milks two cows twice a day, has ten head of horses to work and feed, does equipment repairs, and was, this particular spring, rebuilding the ranch corrals in any spare time. "Milking is my thinking time," she says. "I can get everything all straight in my mind. It's kind of the beginning and end of each day. It's like writing a diary, sitting there milking the cows, but it's relaxing."

We finish feeding the cattle and tagging the newest born calves and are driving back to the ranch with a one-eyed calf for doctoring. She'll bottle-feed him until he's a little older, then she'll put him on skim milk with a protein additive mixed in. "I think women have a little gentler touch with animals," Melody says, while demonstrating her point. "I think they also have more of a mothering instinct so they might be more patient trying to get a calf to nurse or caring for one that's hurt."

As one who cares deeply about the future of ranching in the West she observes,

There are still some good hands [ranching], but precious few, really good hands that live it as their life. Usually you see people work at it five, six, seven years and then get tired of it. It really wasn't as romantic as it was sounding so they're off doing something else. There's more money in the oil patch or whatever their reason.

The freedom with this lifestyle is unlike any other occupation because you have a dual role. On the one hand, you have to be independent and, on the other, you have to be smart enough to blend in with what's happening. You can't be independent to the point of being stupid or bullheaded. I think that's hurting agriculture [and ranching] a lot. Everyone wants to be known by his or her own identity instead of getting together to present a strong unit.

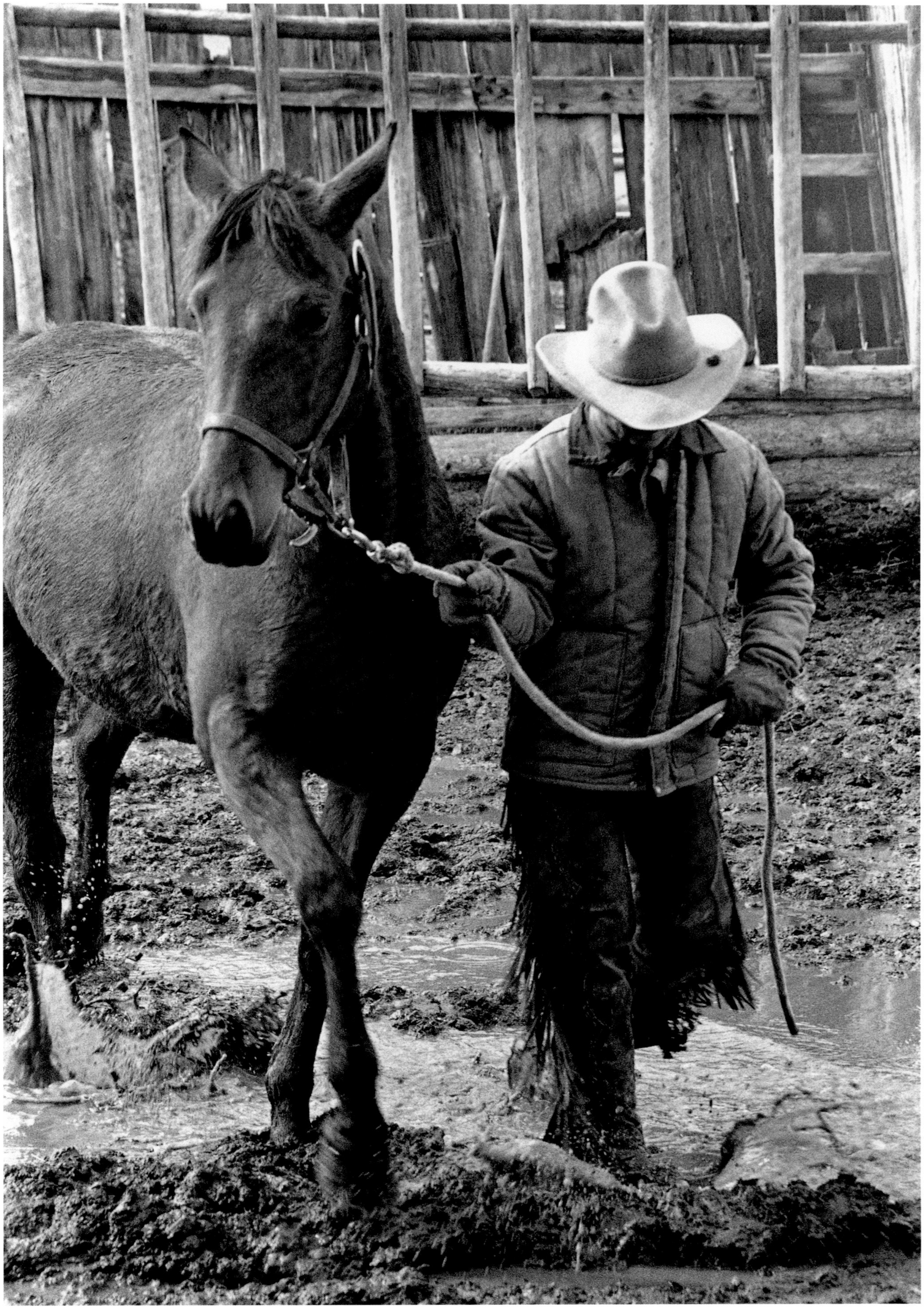

MELODY HARDING, BAR CROSS RANCH
1986

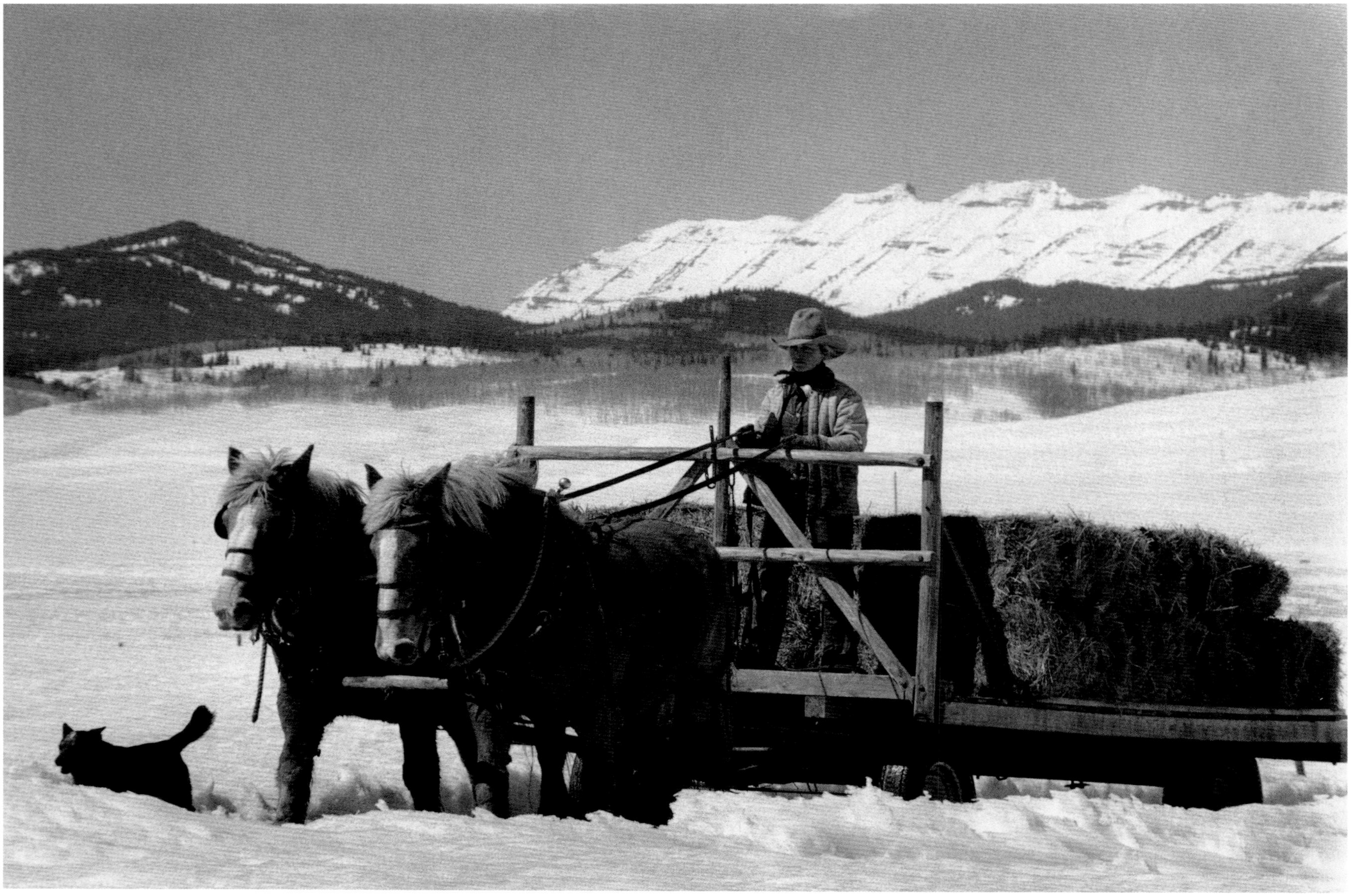

MELODY HARDING, BAR CROSS RANCH
1986

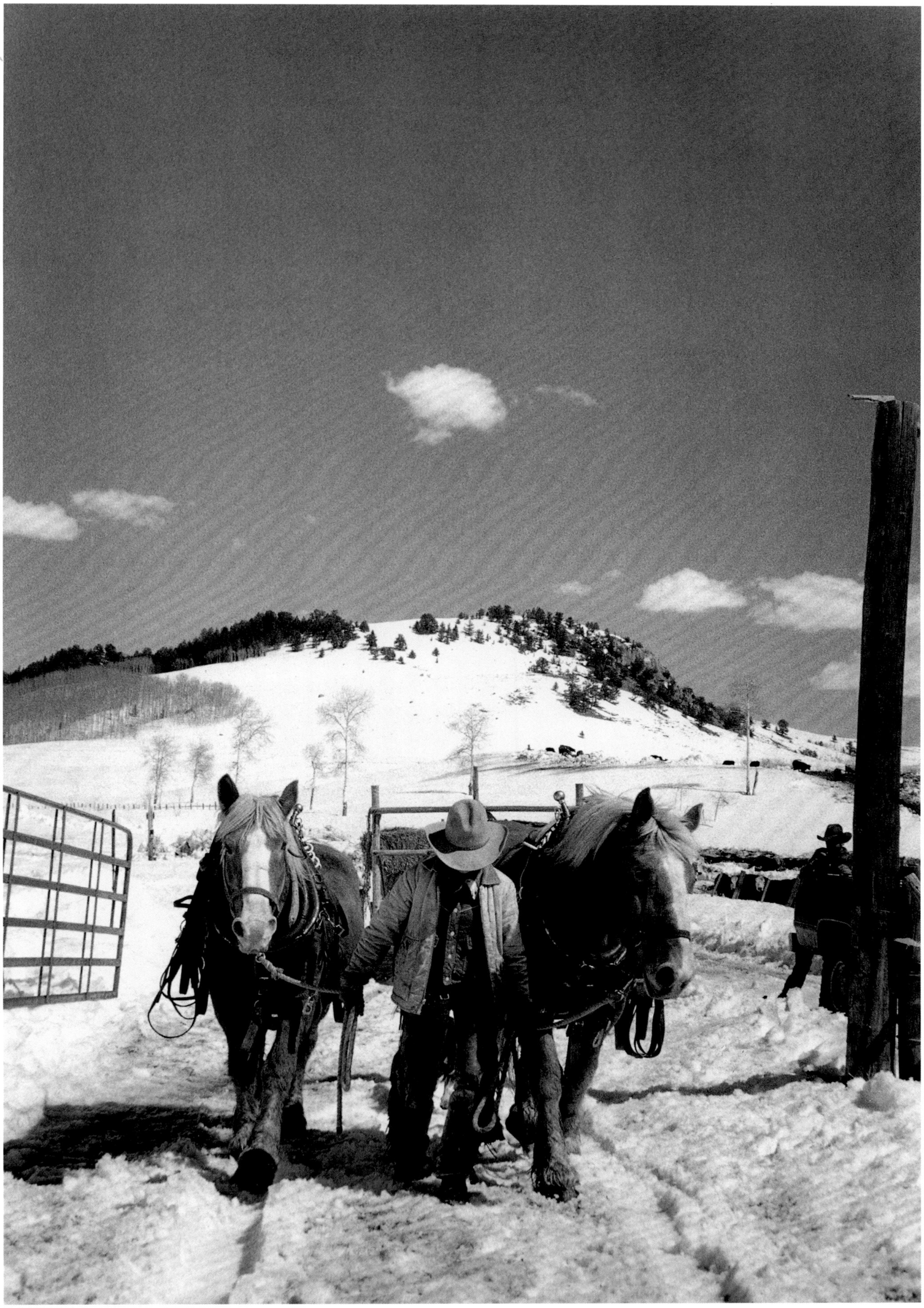

MELODY HARDING, BAR CROSS RANCH
1987

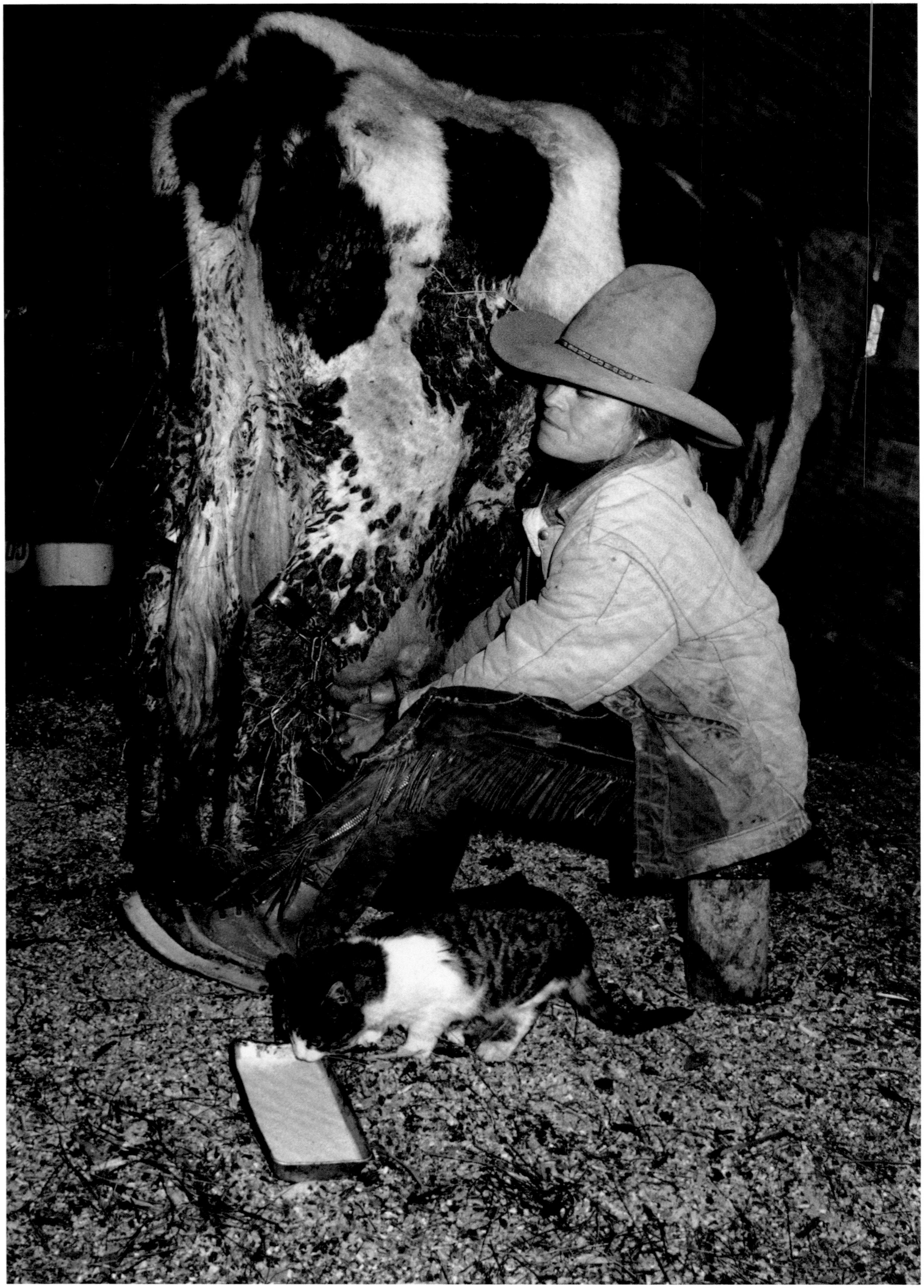

MELODY HARDING, BAR CROSS RANCH
1987

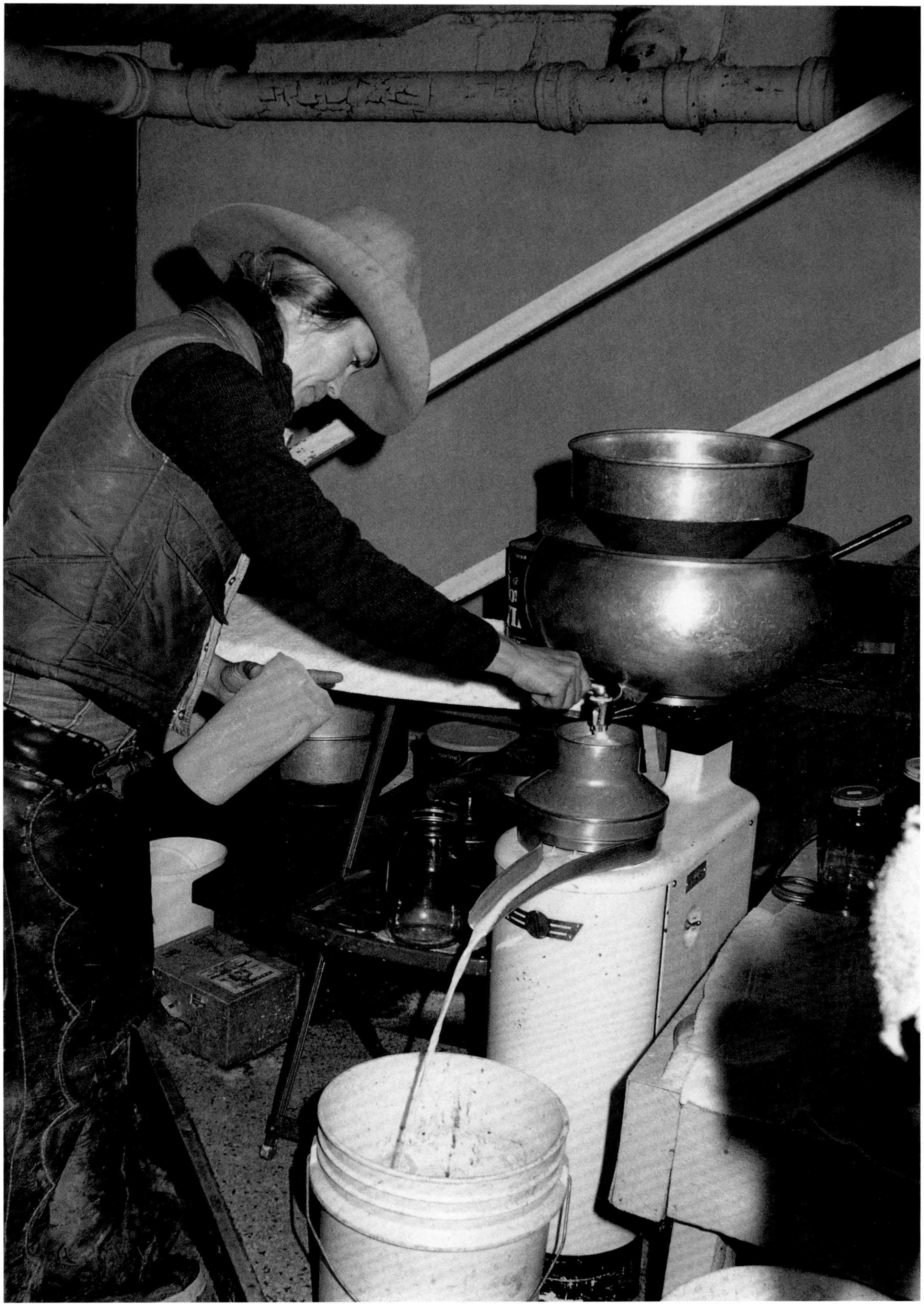

MELODY HARDING, USING THE CREAM SEPARATOR, BAR CROSS RANCH
1987

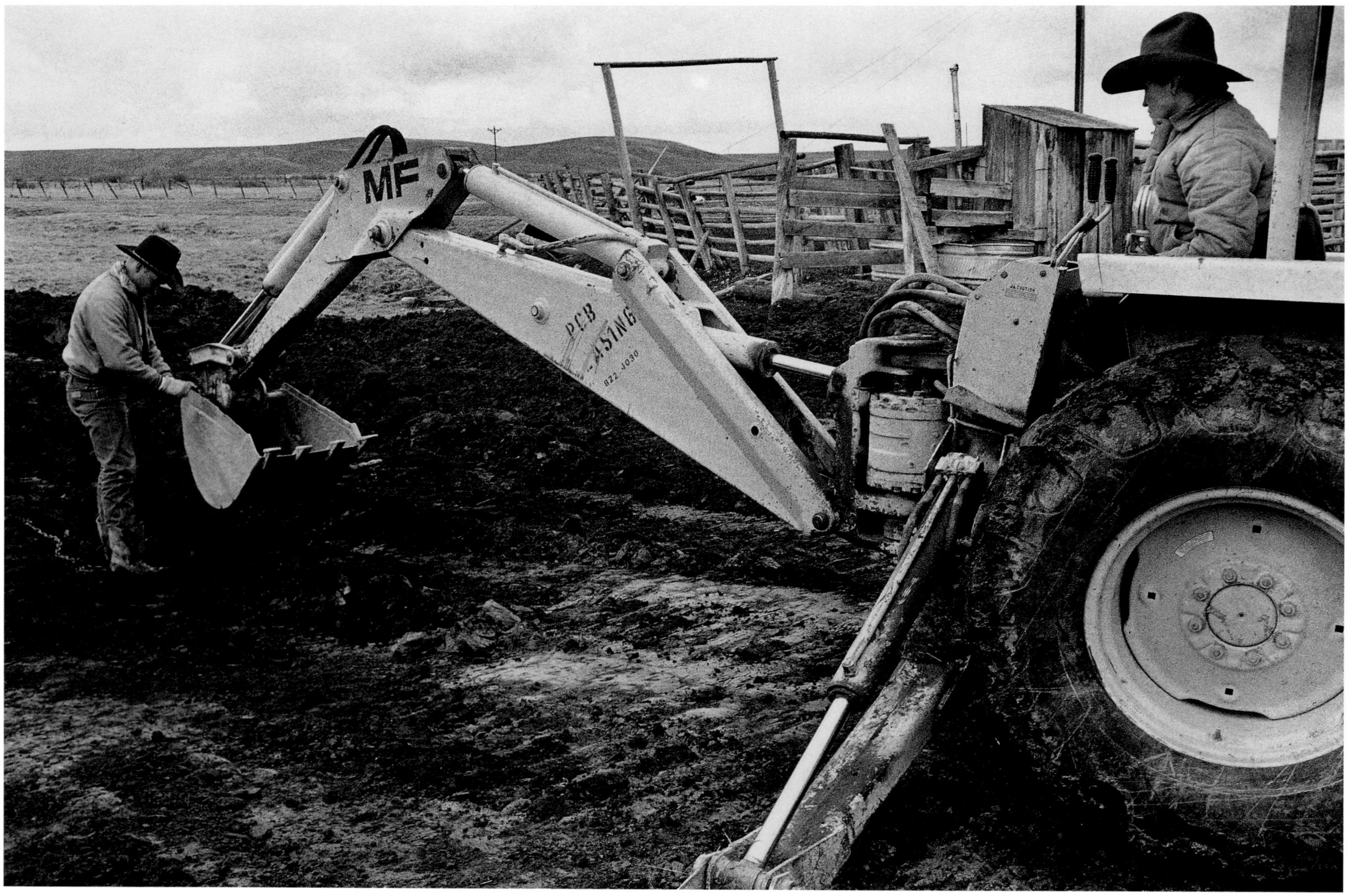

MELODY HARDING, BAR CROSS RANCH
1987

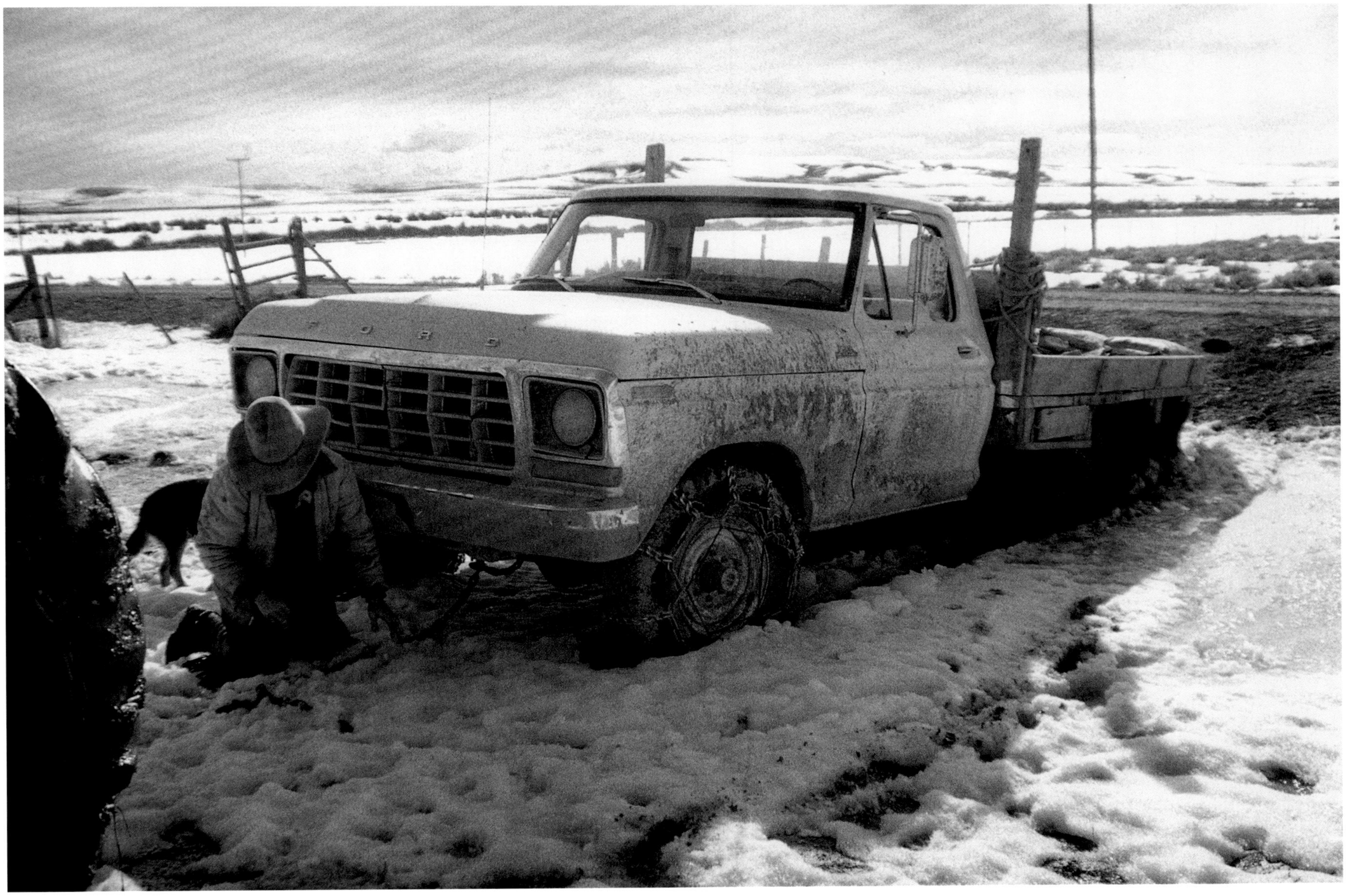

MELODY HARDING, BAR CROSS RANCH
1987

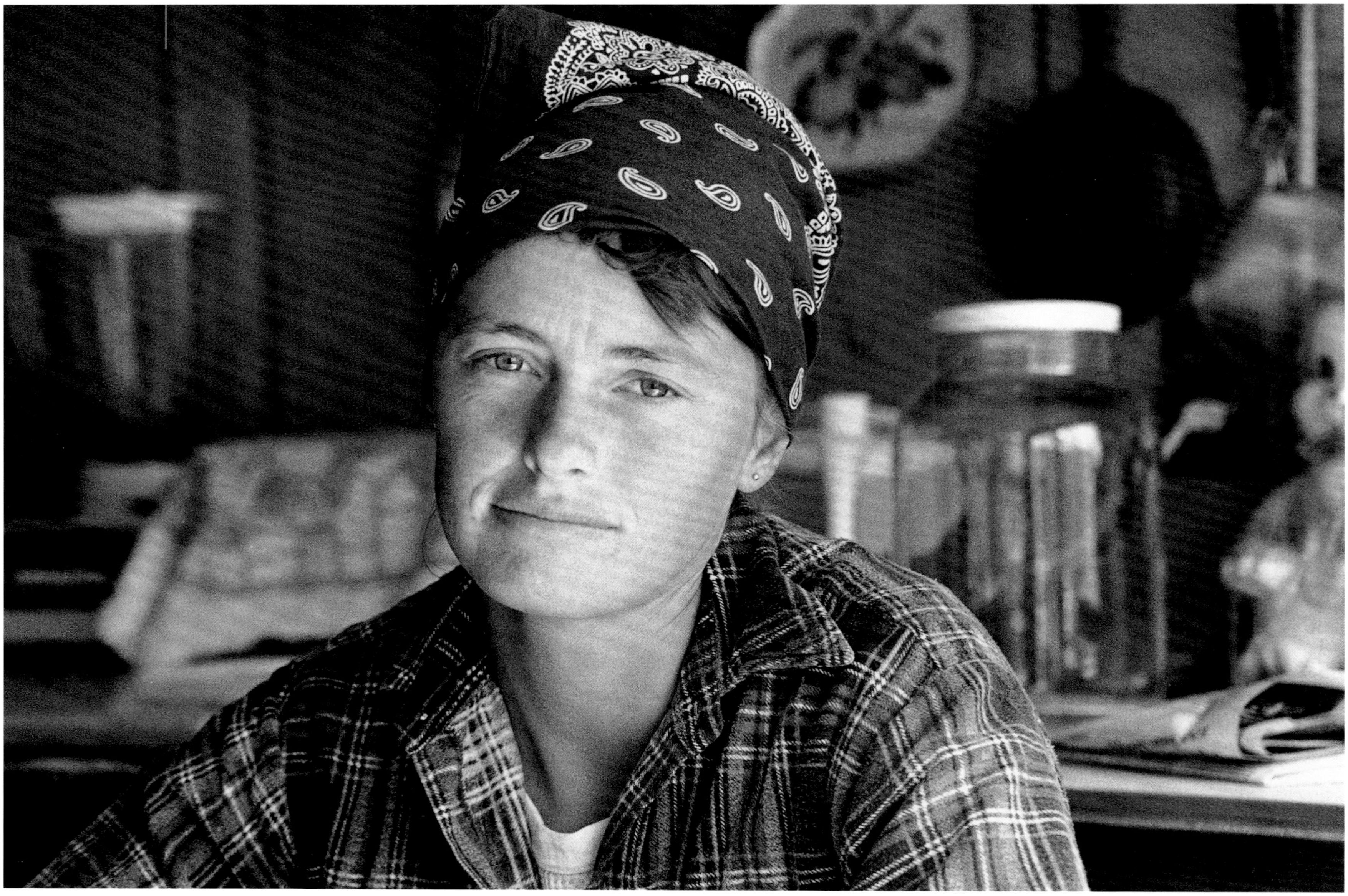

SANDY COOK ROSENLUND, BLUE JAY RANCH
1986

SANDY COOK ROSENLUND

Blue Jay Ranch, Ruby Valley, Nevada

Sandy, slender and quiet, is twenty-six, married with two little boys. A native born and raised Nevadan, she and her husband, Ray, hope one day to buy the Blue Jay Ranch on which they are living and working from her father.

The ranch is 2200 deeded acres plus another 3000 or so acres leased from the BLM. Situated at the south end of the Ruby Valley on the east side of the Ruby Mountains in eastern Nevada, the ranch is remote even by western standards. Sandy acknowledges, "You can't just go out and run down to the store. You have to go seventy miles at least. Town from here, and we're about in the middle of all of the towns, is about seventy to eighty miles from here. If we go to Ely, we only have about twenty miles of dirt road and the rest is highway." More paved roads are going in as gold mining intensifies in the Bald Mountain area.

The distance between the little communities and towns can create a challenge for young people trying to meet, but they do, the way young folks got together a hundred years ago in the ranching West—they go to dances. Ray knew Sandy's father because often his cattle would drift over onto the neighboring ranch where Ray worked and he'd have to drive over to let him know. "Once in a while he'd come over here and visit and stuff," Sandy says, "but it wasn't fancy or formal."

Much of the Ruby Valley is being developed by the government and real estate interests. The sagebrush has been cleared—ranchers *would be rich*, truly, if someone found a use for sagebrush—irrigation systems have been installed, and farmers are raising hay commercially. According to Sandy, the federal government encourages people through the Desert Land Entries program. It will sell you 360 acres for $96, provided that you can put it into production within two years. Then they will give you a deed for that land. You have to clear the land of sagebrush, drill your wells, put in your irrigation, and start production within that time period or you don't get the land. "All those improvements are pretty costly," Sandy says. "We tried it here, but we just couldn't handle the cost of the improvements within that time frame."

However, Sandy and Ray are satisfied with their ranch: "We have four good springs, all deeded, on this mountain which pretty much takes care of this land. Because we don't put up or raise hay, we have to buy some to feed our weaner calves. We have enough winter range to run the cows out there, so we don't need much hay." And they cut, split, and sell firewood to help bring in extra cash when beef prices are down. This ranching couple is just breaking even.

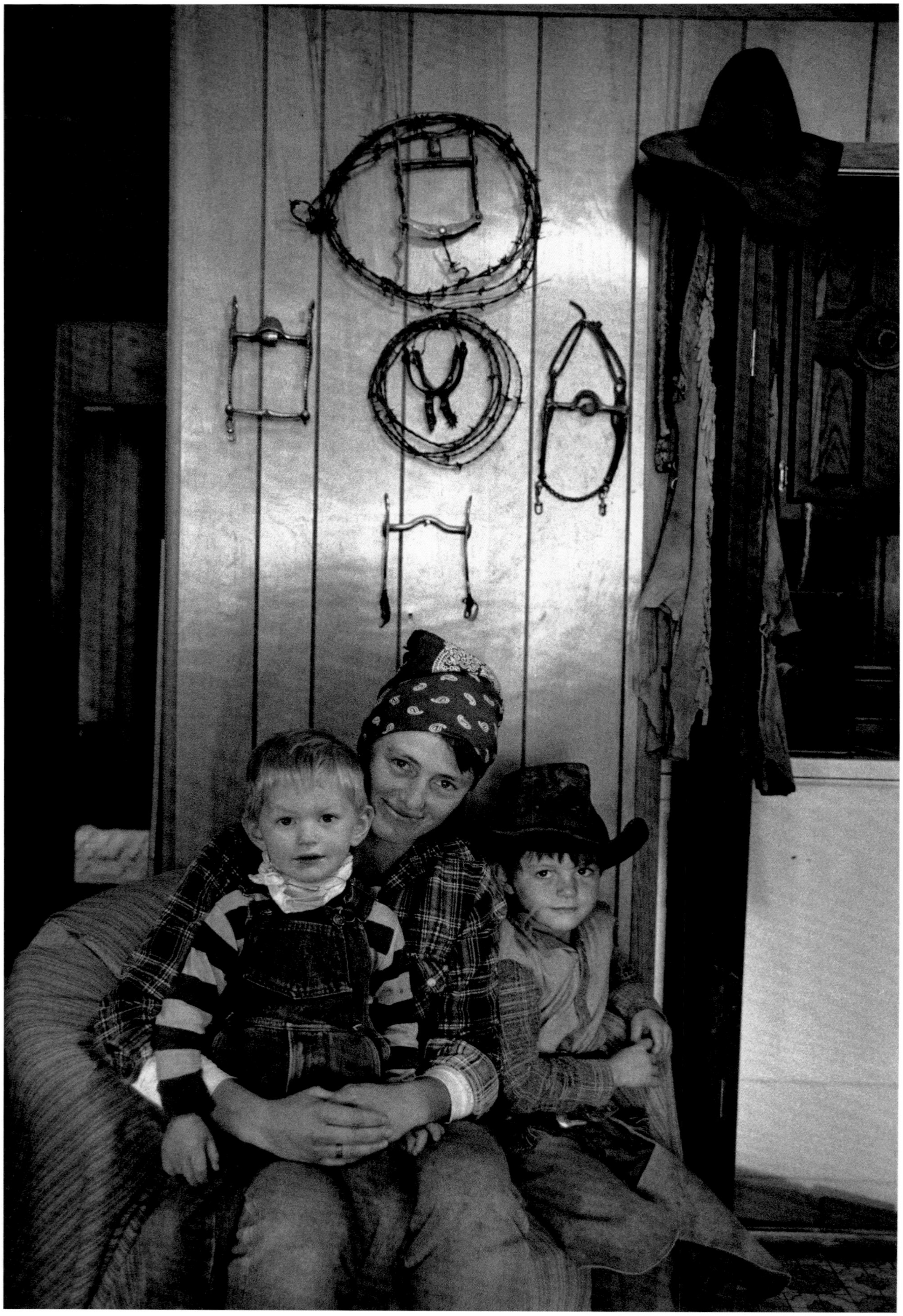

SANDY COOK ROSENLUND AND SONS, BLUE JAY RANCH
1986

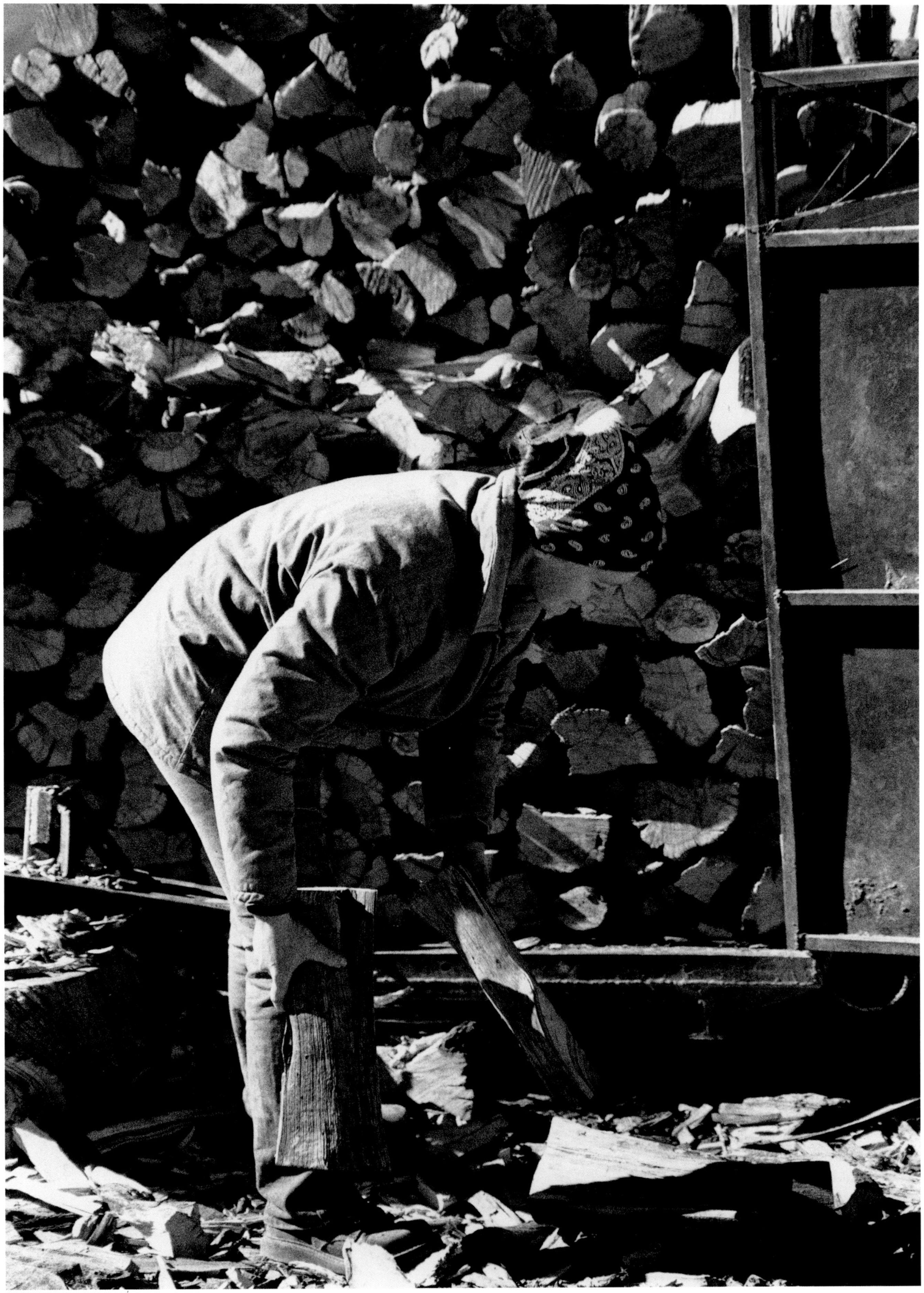

SANDY COOK ROSENLUND, BLUE JAY RANCH
1986

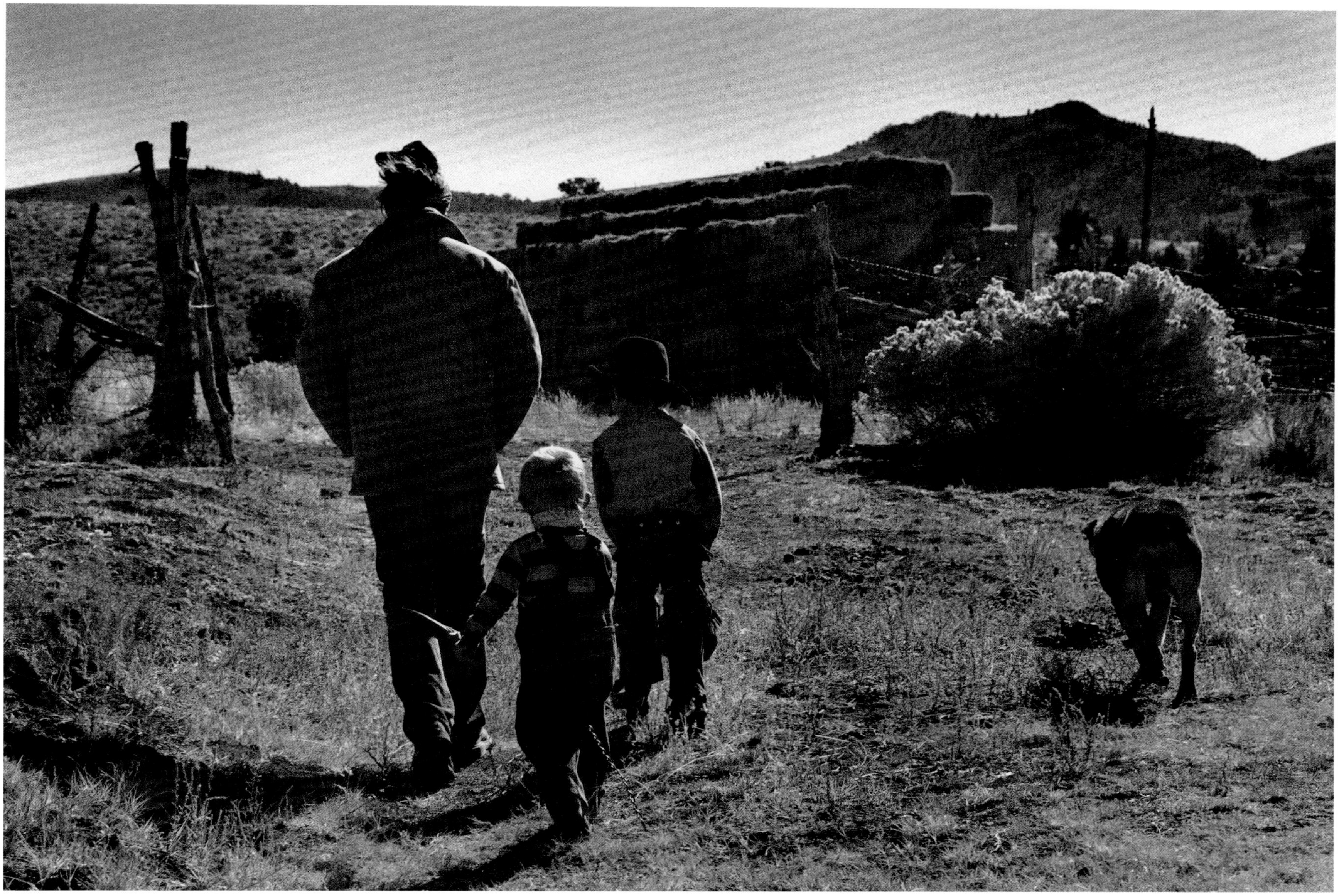

SANDY COOK ROSENLUND AND SONS, BLUE JAY RANCH
1986

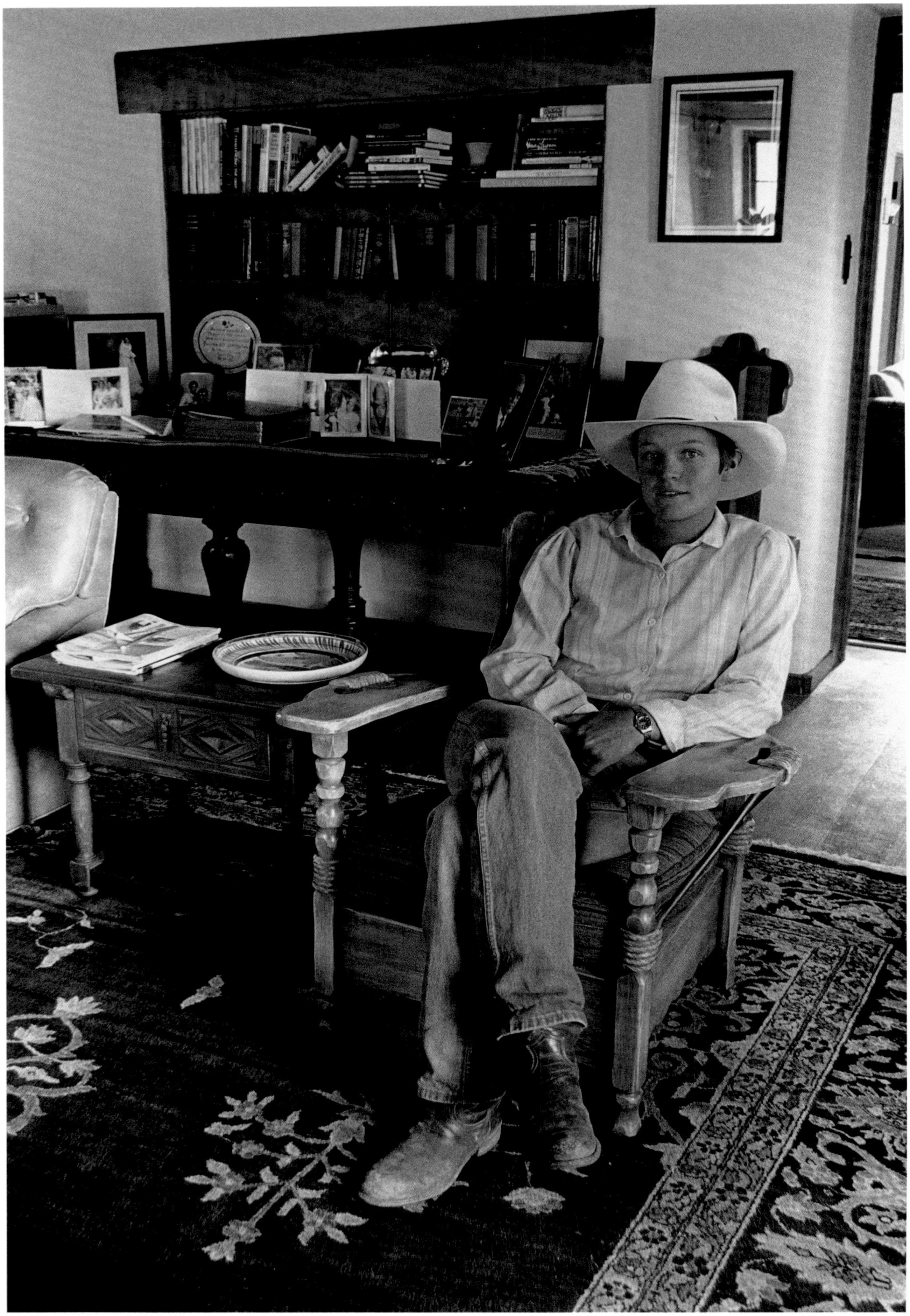

KIM DAVIS BARMANN, CS RANCH
1986

Davis Family

CS Ranch, Cimarron, New Mexico

Sixty-four-year-old Linda Mitchell Davis with her husband, Les, and their six children, five of whom are married, own and run the historic CS Ranch outside Cimarron, New Mexico's largest family-operated ranch. There along the Cimarron River, which cuts through the green and grassy plain that stretches out from the base of the Sangre de Cristo Mountains, lies the ranch headquarters. There is a lot of history about this ranch and where it is situated, straddling the Cimarron cutoff of the Santa Fe Trail. It is good grass country, with good water and enough wind to move the winter's snows from the flatlands so the cattle have fairly easy grazing. The peaceful look of the well-kept ranch buildings shaded by lovely old cottonwoods can be deceiving, for no one at this ranch sits around on the front porch and "takes the view."

Les Davis's grandfather, the famous Territorial lawyer and paleontologist Frank Springer, went to New Mexico in 1873 and began to accumulate the land that would eventually make up this 250,000-acre family-owned and -operated ranch. Springer's brother Charles was one of the early managers. His initials, CS, were first recorded back in 1880 as the ranch brand.

Linda was raised not far from Cimarron on the Tecosquite Ranch at Albert, New Mexico, established by her great-grandfather in 1878. Her father, Albert K. Mitchell, at one time managed the renowned Bell Ranch near Albert, in addition to running the family ranch of some 200,000 acres. He was, among many other things, one of the founders of the National Cowboy Hall of Fame and Western Heritage Center in Oklahoma City.

It was at the Mitchell ranch that pre-med student Les Davis met his future father-in-law, Albert Mitchell. In 1941 his uncle, Ed Springer, who wanted Les to *stay on the ranch*, brought the recent Dartmouth graduate to meet Mitchell and "see how a *real* outfit is run." There he met the young Linda Davis. "I never forgot him," she says with a smile, "even though I was only eleven and he was a debonair twenty-two." That day, Linda recalls, "Dad [who was by then a widower] told me to gather certain cattle and take them to the ranch headquarters, and he said, 'I want a good dinner, too.' So I gathered the cattle, cooked dinner, and did everything else too. Les doesn't even remember!" They eventually married in 1953 and have lived on the CS Ranch ever since.

Linda, a warm, bright, and attractive woman, is constantly on the move and involved in all areas of ranch life. At 4 a.m. she is organizing for one of the days of spring branding. An avid hostess, she plans and prepares banquets for such social affairs as the New Mexico Cattle Growers annual meeting. Vaccinating calves every day for two weeks of each spring is another activity in her busy life. She has been named "Cattleman of the Year" in New Mexico and the *Progressive Farmer*'s "Woman of the Year in Service to Southwest Agriculture," and has received numerous other well-deserved awards in the beef industry.

Dad was a widower from 1934 to 1957 so he was great about taking me along all the time. As the oldest I reaped the benefit not only of being his friend but also his confidante. And I had the advantage of being a girl. I was very close to my dad. Les is the businessman. I'm the cattlewoman. I was raised in the saddle so I have no excuse not to be a good cowboy. And I think that the fact that I was raised around the Spanish people and their old-time families who had worked and lived on the ranch for generations made me work harder at being a real go-getter and top-notch hand. But it was sad to see how those old Spanish families would react when the oldest child was a girl. They would shake their heads and say, "Too bad, because she likes it." But I was a girl, the oldest, and I had a natural aptitude.

In part I was a tomboy, too, always was. I got to go with the men if I behaved myself and didn't scream or wail. The net result was that I probably turned out to be more of a stoic because I was the only girl and I wanted *to go with the men. I'm not afraid of snakes and such. Of course, if you're a girl, you always know that you can do the cooking and probably* will. *You can ride and cook, too! I remember hearing so often as a youngster at the Bell Ranch, "Well, Linda's along, so* she *can fix the white shirts for Saturday night." Of course I was my dad's hostess and housekeeper. Later I planned all the meals.*

Keeping the family ranching tradition together is important to the Davis family. The youngest daughter, Kim Davis Barmann, is quiet and rather shy. A pretty young woman of thirty-four, she and her husband, Eric, have two daughters and live on the Clayton Place division of the ranch, outside of Springer. Kim echoes her mother's concern:

My biggest worry is keeping the ranch together. It is going to be a chore and we're going to really have to work at it. Times are getting tougher for ranchers and we need to improve so that we won't lose any of the ranch. Big families can have big problems. Even though we [the children] are close, it is harder now because most of us have spouses and we have to deal with their feelings, too. We're working it out, though.

We had to learn to become more communicative. About eight years ago we began having regular meetings. We had to open up so we all would feel comfortable stating problems as we saw them instead of keeping things inside and trying to be buddy-buddy.

I admire my parents tremendously for how they've raised us six kids. I had so many classmates that had problems with drugs. I never remember Mom or Daddy lecturing us about the dangers of drugs. They never laid a hand on us. We just knew *that we'd better snap into line. I was the baby, so the rest of the kids just kind of "loose herded" me around. It was a good childhood.*

Kim figures the CS runs around 160 head of horses; she wants to develop the horse breeding and training part of the ranch. "On the average, we've got about seven to ten two-year-old colts to break each year. I'll be training and finishing them mainly for the team-roping market. I'm getting a much better quality colt every year in terms of brains, savvy, and confirmation."

Kim's older sister, Julia, is an attractive and engaging young woman. She looks younger than her thirty-nine years. An attorney, she has been married to fellow lawyer David Stafford for ten years. "I don't handle all of the ranch's legal matters. There is a lawyer who has been handling its legal work for years."

She, like many of the other children, went away for her education, because Linda and Les thought "we kids would be so isolated if we lived out on the ranch for the rest of our lives after getting a local education. They were afraid we wouldn't have a clue about anything else in the world. But they never told us we had to come back to the ranch after we got our educations. We had a choice—they really wanted us to do what we wanted to do with our lives."

After attending Cornell University and graduating from the University of New Mexico, Julia returned to the ranch to take stock. In law school she concentrated on natural resource courses, aware of the critical role that water especially plays in New Mexico.

I worry a lot about the ranch. These days are times of real change for the ranching business. Now we can know exactly what it costs us to raise a cow, because we are keeping track of every tiny expenditure on the ranch related to raising cattle. And there are the family differences. It's touchy. The Davis family that was once just one very strong family is now seven *very strong families.*

I'm torn between my professional life and the ranch. Sometimes I do things that I enjoy a lot more than doing things I should be doing. I have every intention of being a top-notch trial lawyer, eventually. But I also want to see the ranch—such an amazing place—be the best of everything it can be.

David made an interesting observation. He said, "What is amazing about rural people is that they can fix things, any thing." It's true. If something breaks down, you figure out what is wrong with it and then you find or make something that does the same thing. Rural people are resourceful probably because it is just too long a trip to town to the store. And, new parts cost a lot of money. Ranching is such a different way of life that it is almost impossible to explain to somebody. Take the matter of space. It is overwhelming to most city people. How do we explain the whole kind of openness and freedom that we have? And how do I explain that we're raising these animals, knowing that they're going to be slaughtered for beef, yet we are out there in some terrible driving storm trying to save them and do something for them, feed them or bring them in. How do I explain to somebody that no, I don't get bored or lonely, nor am I frightened about being "out there," all alone. The pull, the lure, the attraction, the sense of responsibility for the ranch will always be with me no matter who I'm with or where I am.

A dark-haired, bright-eyed, pretty girl from Georgia named Mary Bailey married the oldest son, Warren, forty-one, who manages the Crow Creek division of the ranch. Mary and her father had been coming out to Raton, New Mexico, for a number of years to buy quarter horses. During those trips they had become good friends with the Davis family. "Dad came out here to pick up a stallion he was leasing. I had been showing horses for him since I graduated from high school, and I didn't want to leave Georgia. I'm a real homebody. But I did. The first time I met Warren I knew right then that I would marry him."

Mary is happily settled in with Warren, and their children, Walter and Sarah. When she moved out to the ranch she brought her favorite horse, Flash, a nineteen-year-old handsome Palomino gelding, with her. "He's my rope horse, my pride and joy. Jiggs Porter [ranch foreman] taught me how to rope. I'd like to get as good as Kim is."

Mary has never regretted the move away from Georgia, nor questioned her destiny in becoming part of the Davis family.

I've never gotten homesick. Never, not once. My best friend in Georgia came to visit. She thought she was in the middle of nowhere, then the phone went out! Whenever we get rain or snow or the wind blows too hard, the phone goes out. I couldn't believe that there was so much difference between the two of us now. Actually, I've always liked being alone, so being here by myself doesn't bother me at all. I hate to wish the years away when the kids are so young, but I can't wait 'til they are all in school and I can get back to helping with the ranch work. I want to be out riding every day. I'm terrible, but that's all I want.

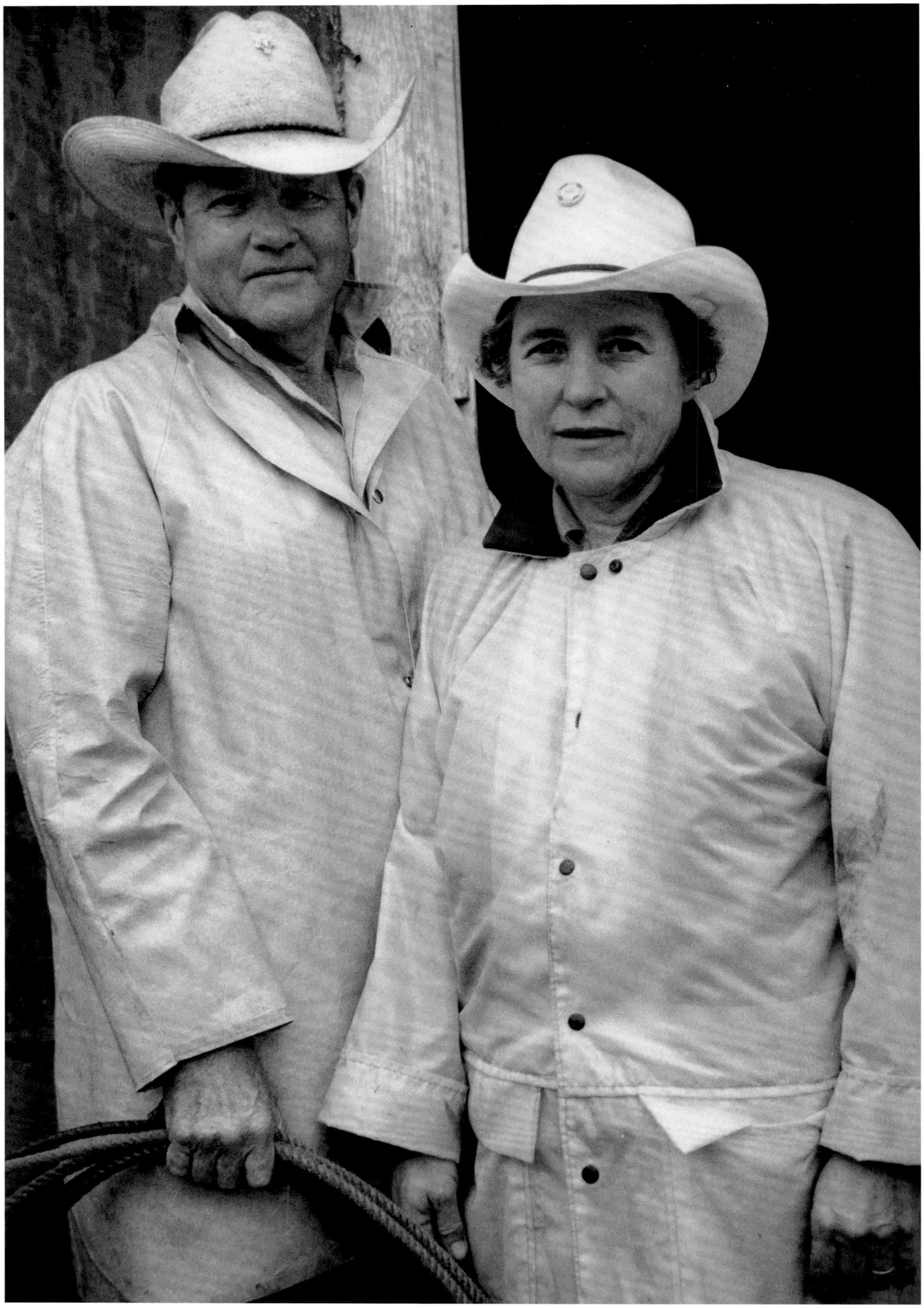

LINDA AND LES DAVIS, CS RANCH
1987

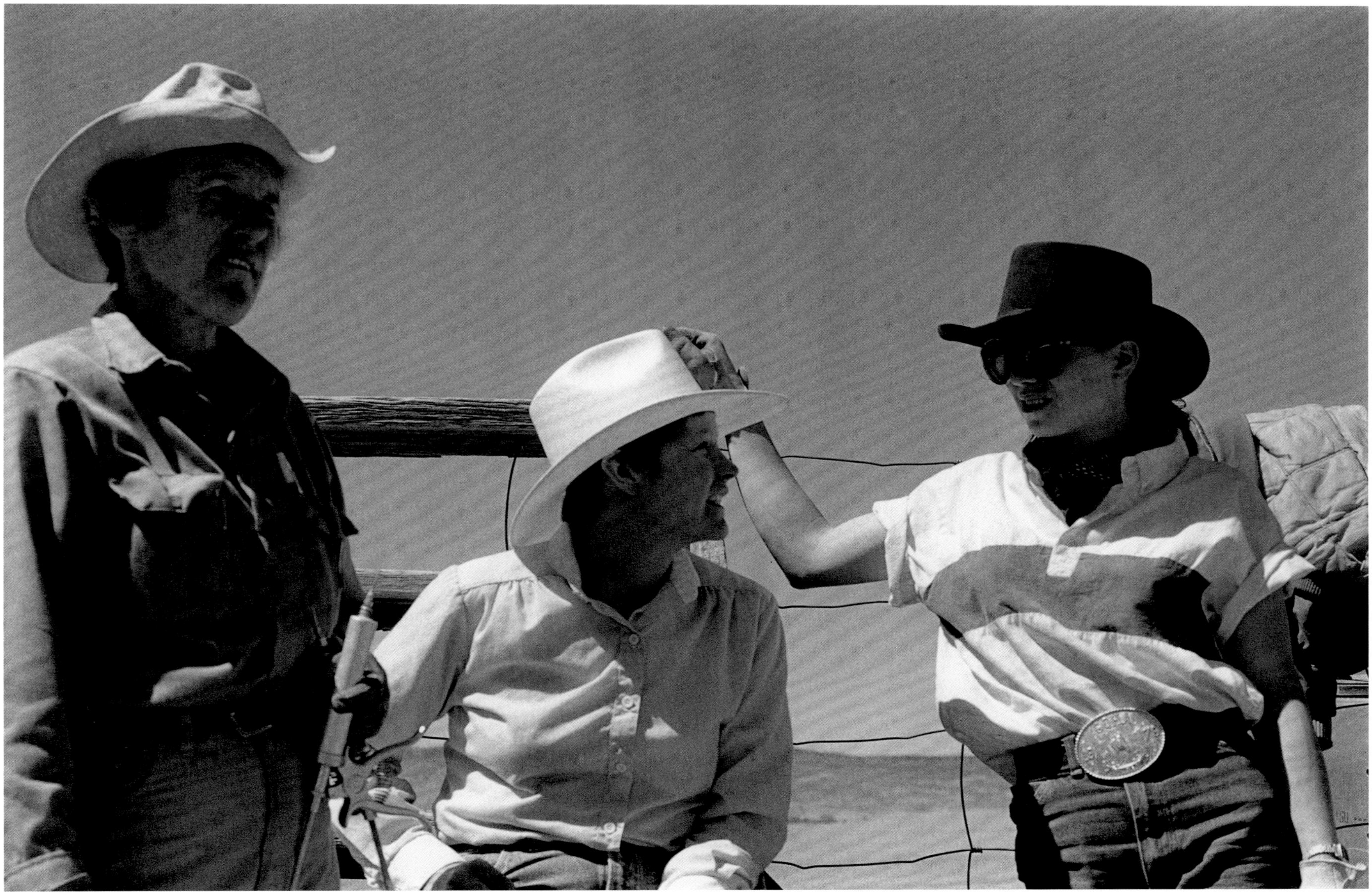

LINDA MITCHELL DAVIS, KIM DAVIS BARMANN, AND MARY BAILEY DAVIS, CS RANCH
1986

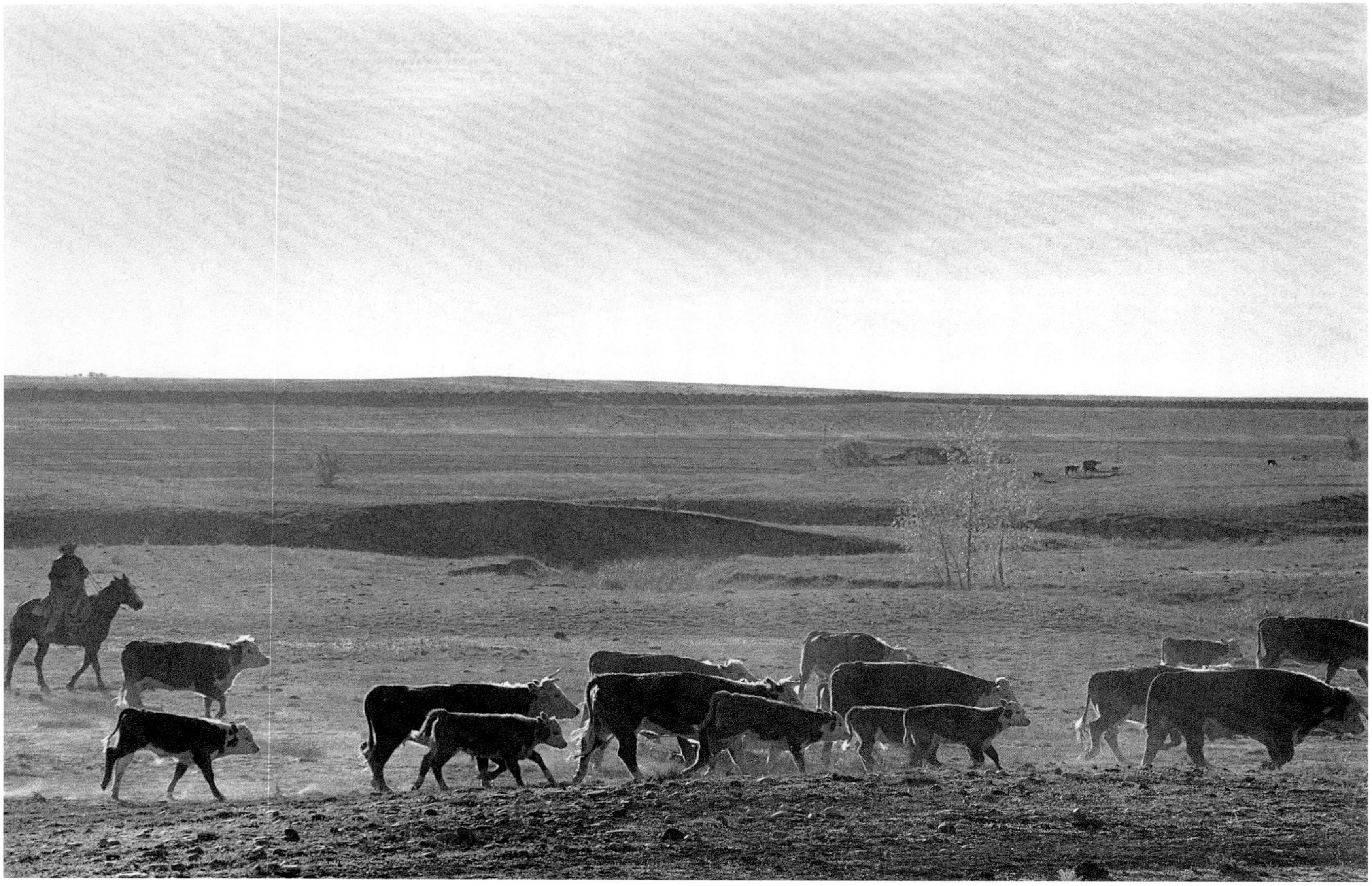

LINDA MITCHELL DAVIS, CS RANCH
1986

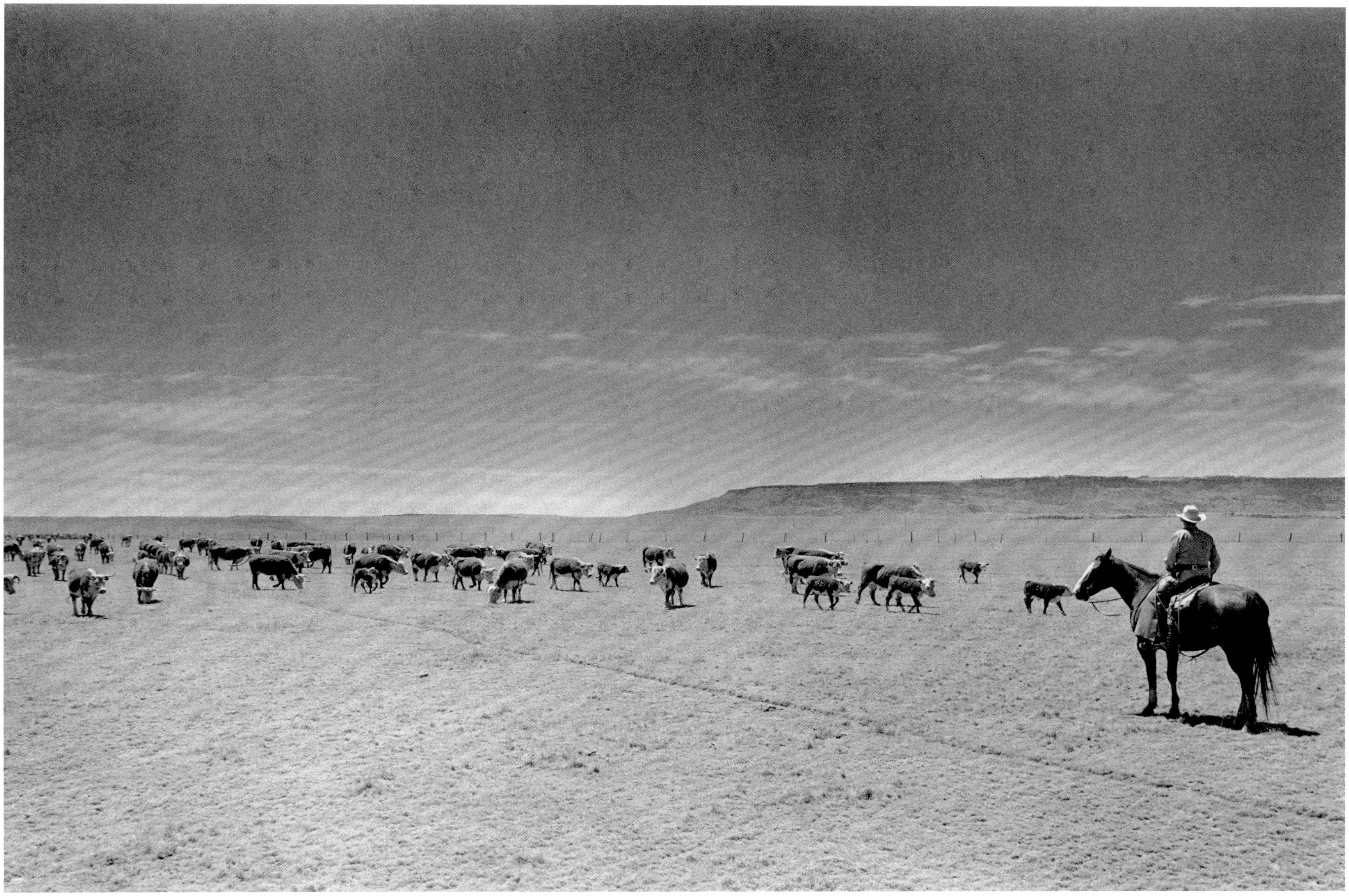

LINDA MITCHELL DAVIS, CS RANCH
1986

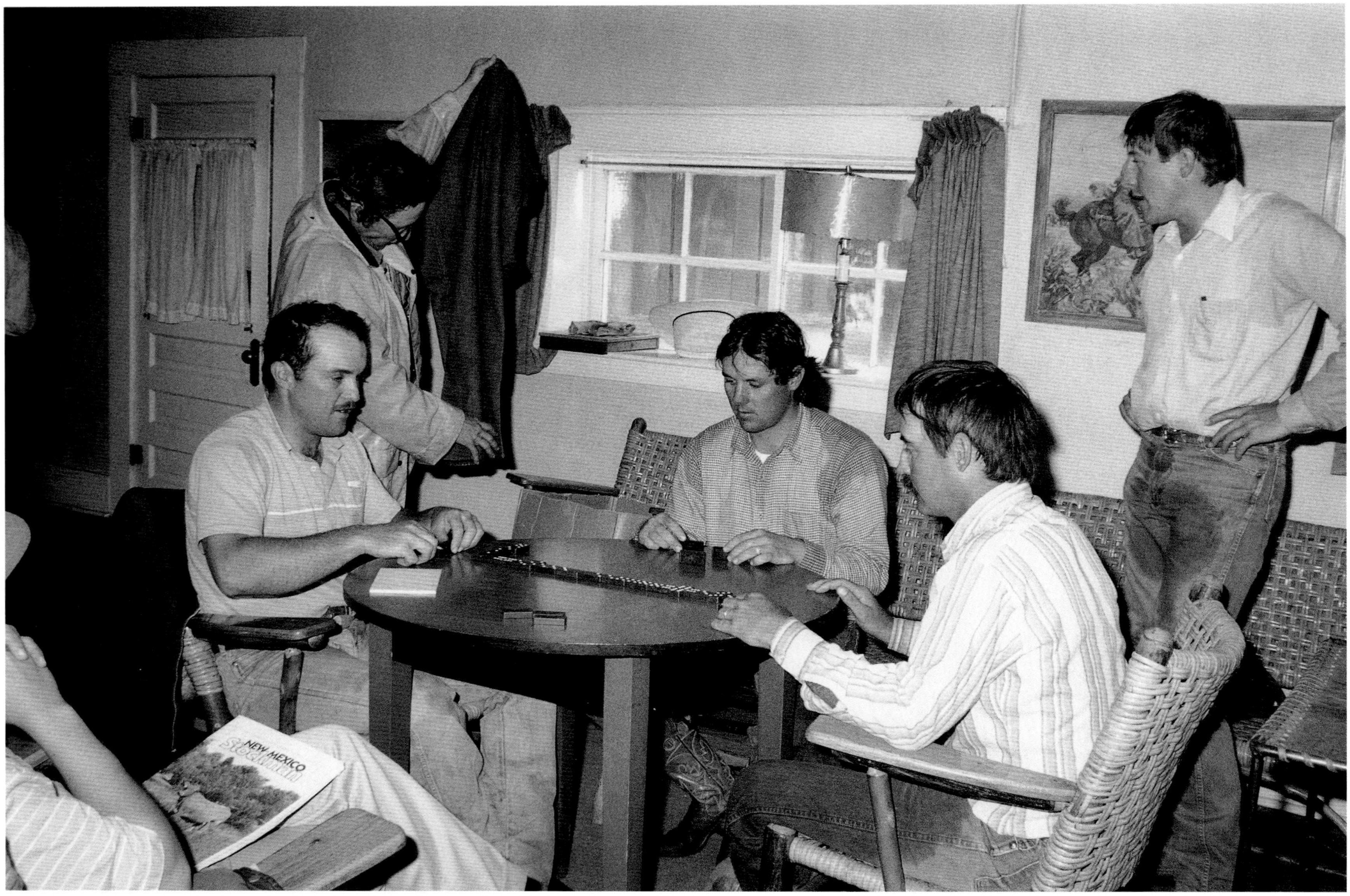

DAVIS FAMILY, AFTER GATHERING CATTLE IN A RAIN STORM, CS RANCH
1987

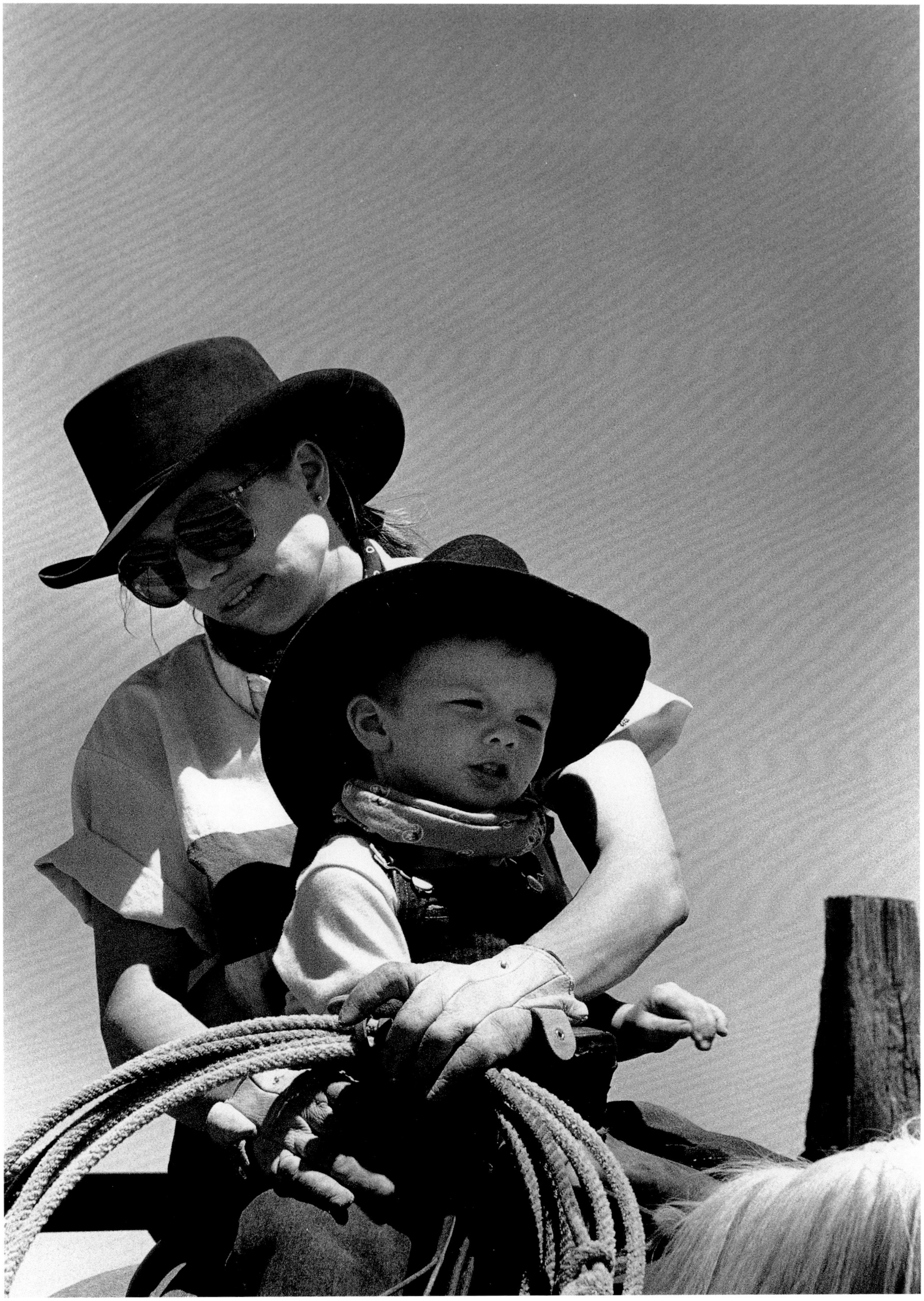

MARY BAILEY DAVIS AND SON, WALTER, CS RANCH
1986

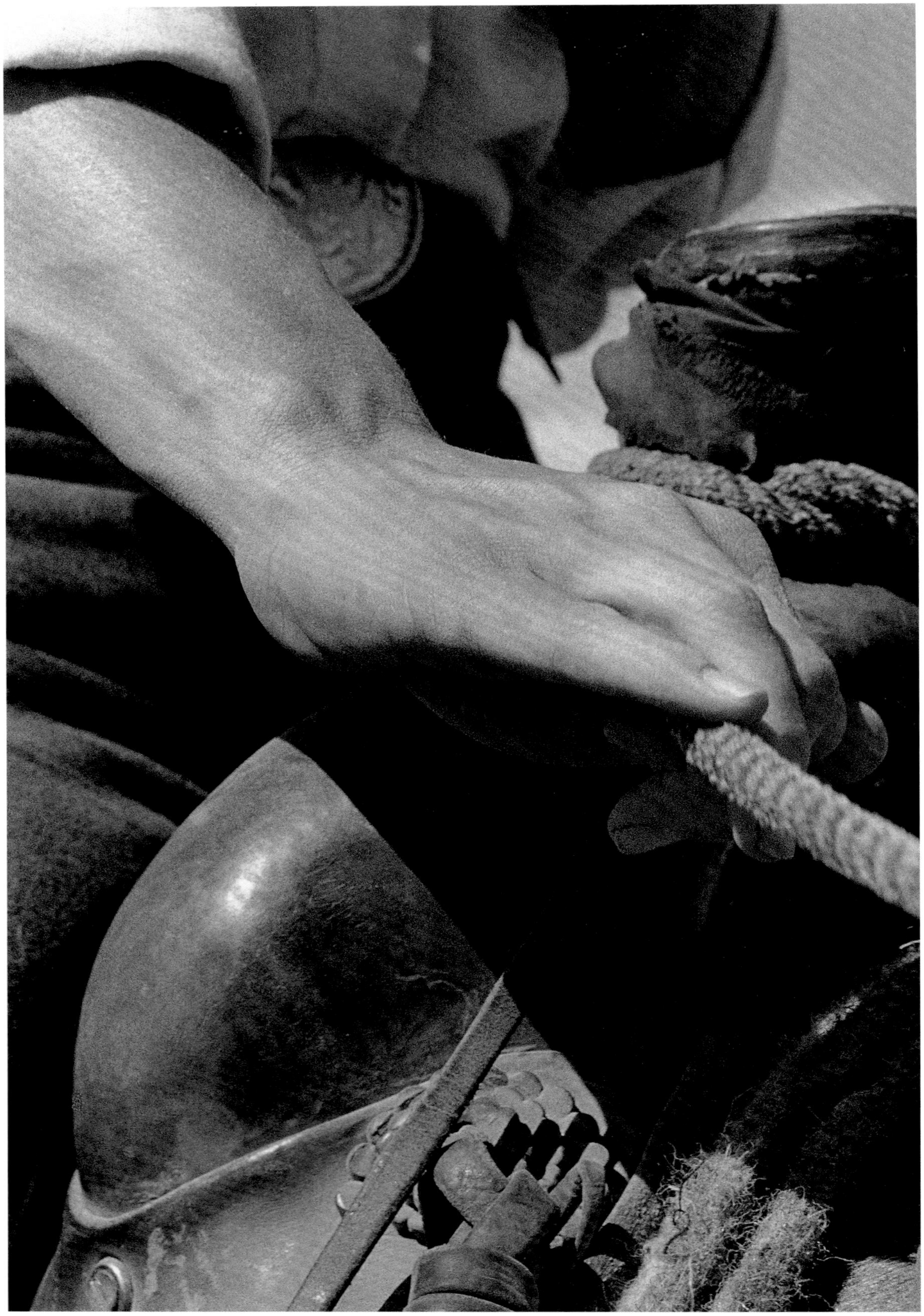

MARY BAILEY DAVIS, CS RANCH
1986

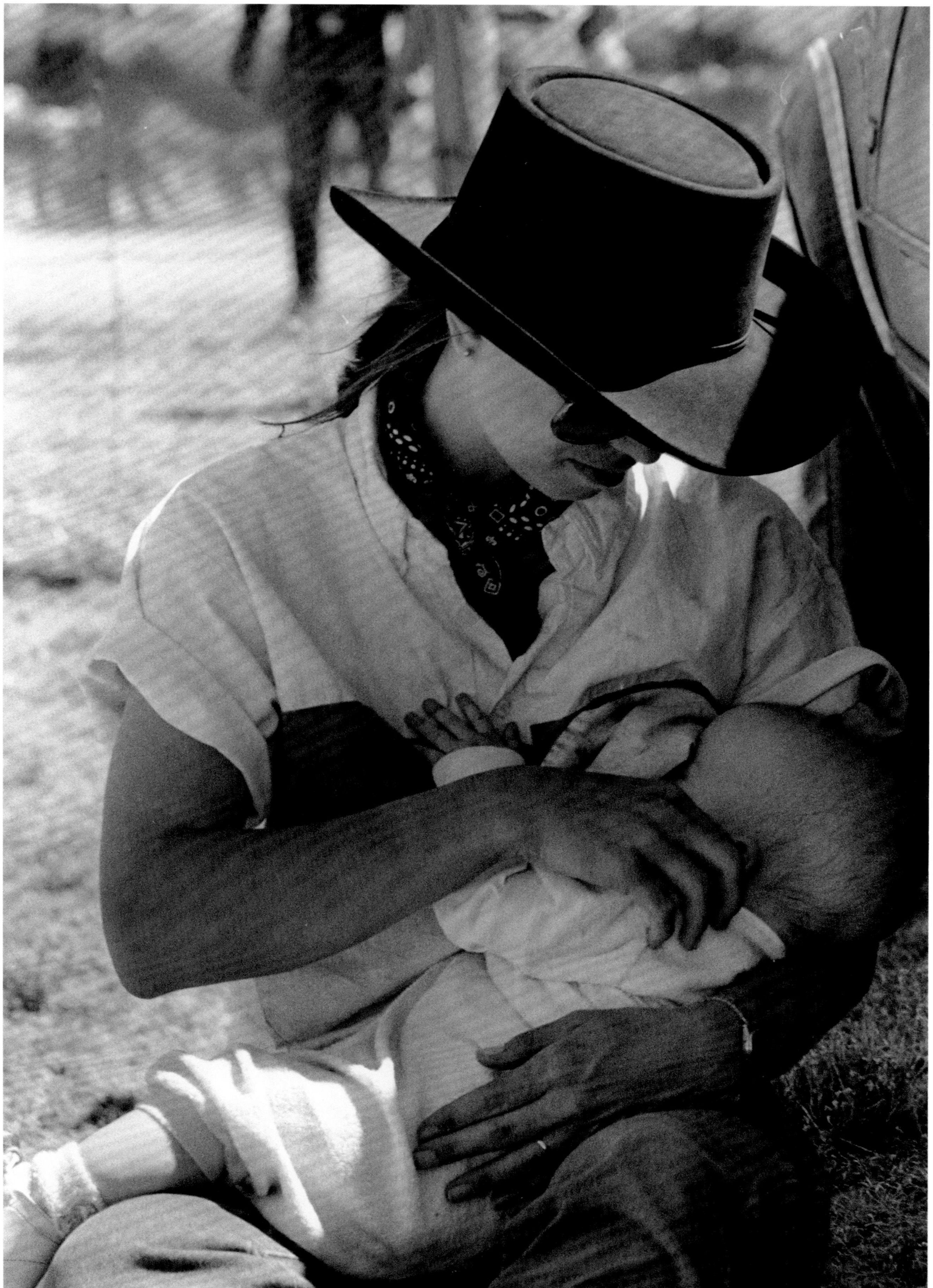

MARY BAILEY DAVIS AND DAUGHTER, SARAH, CS RANCH
1986

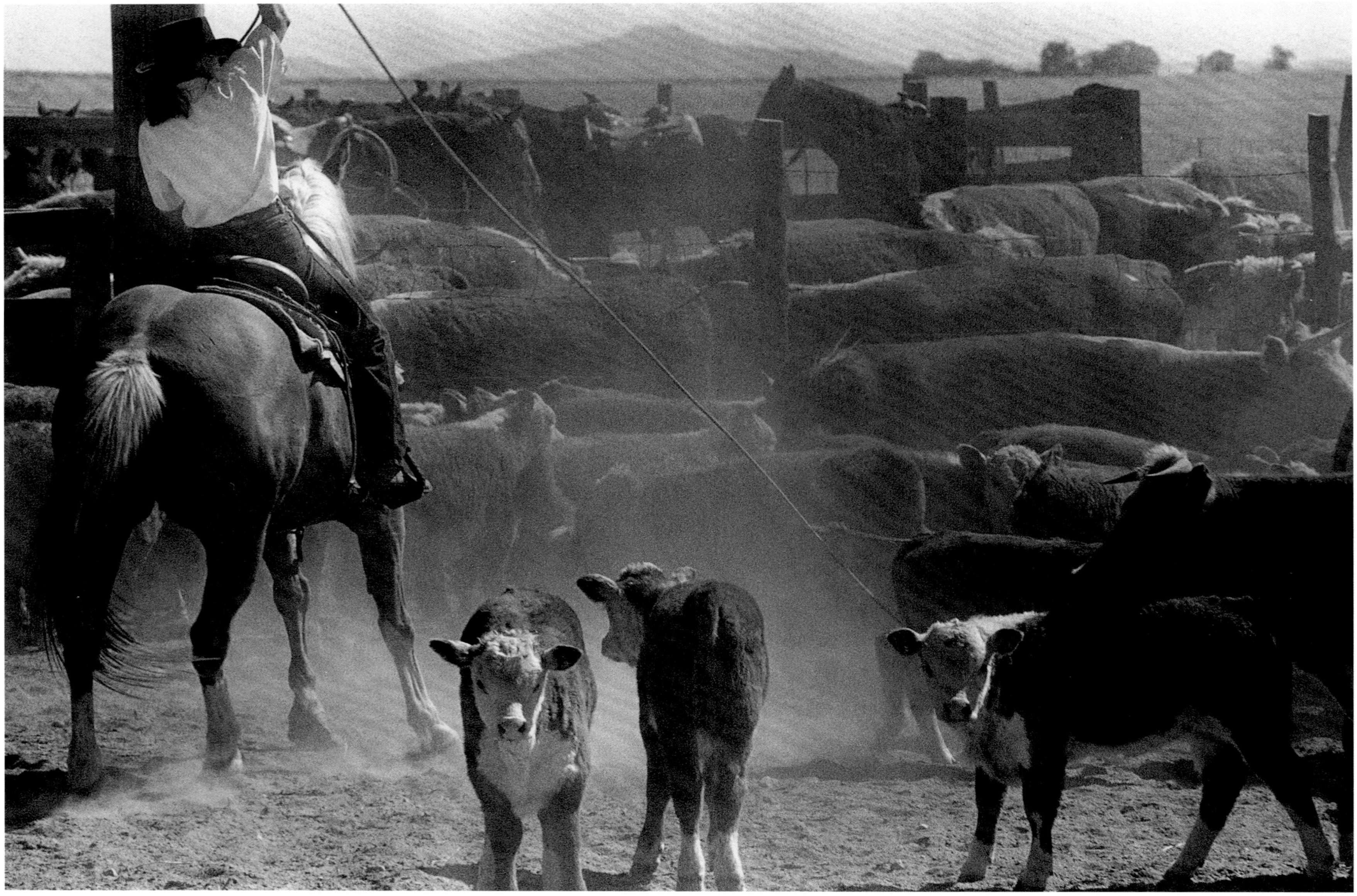

MARY BAILEY DAVIS, CS RANCH
1986

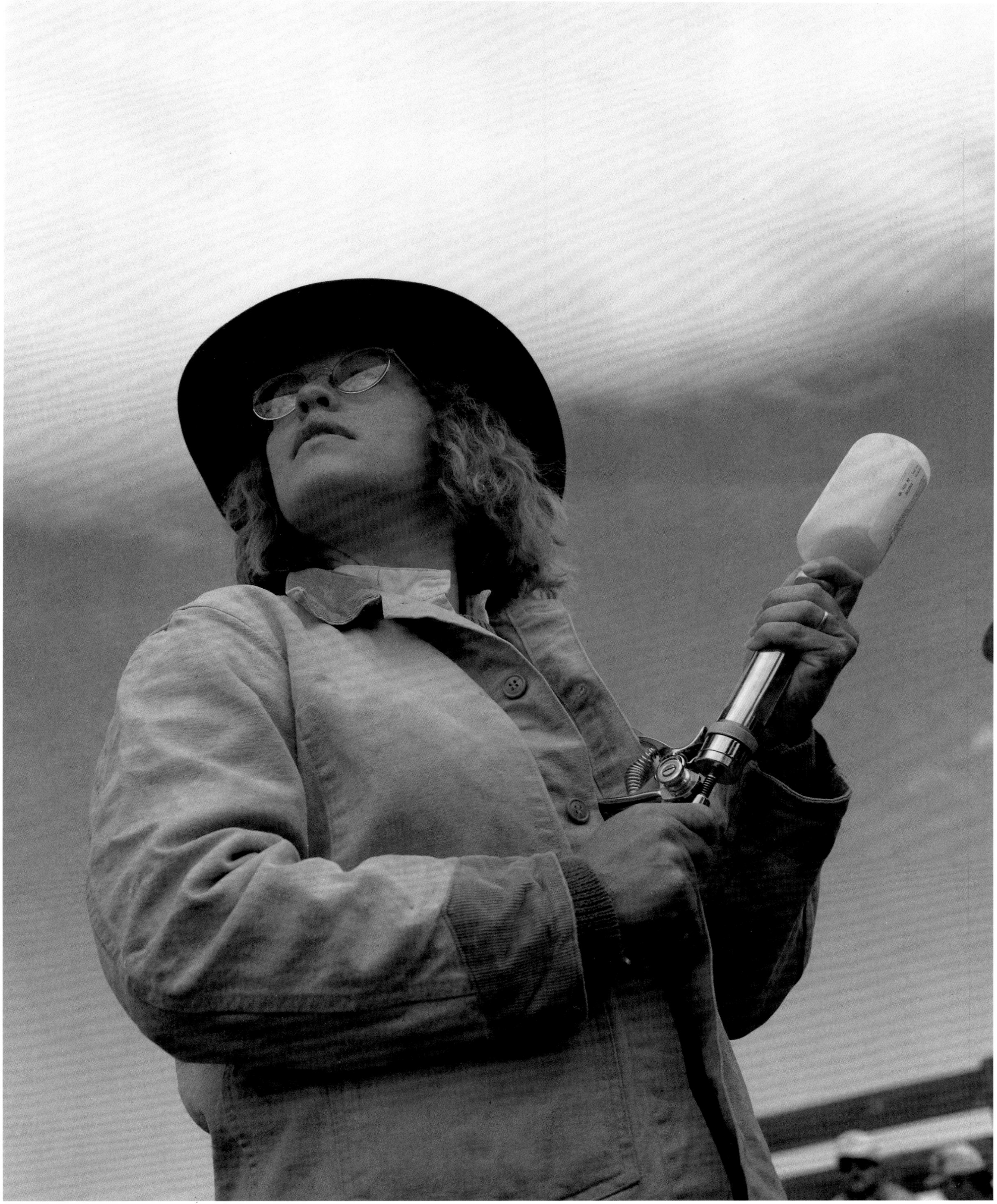

JULIA DAVIS STAFFORD, CS RANCH
1987

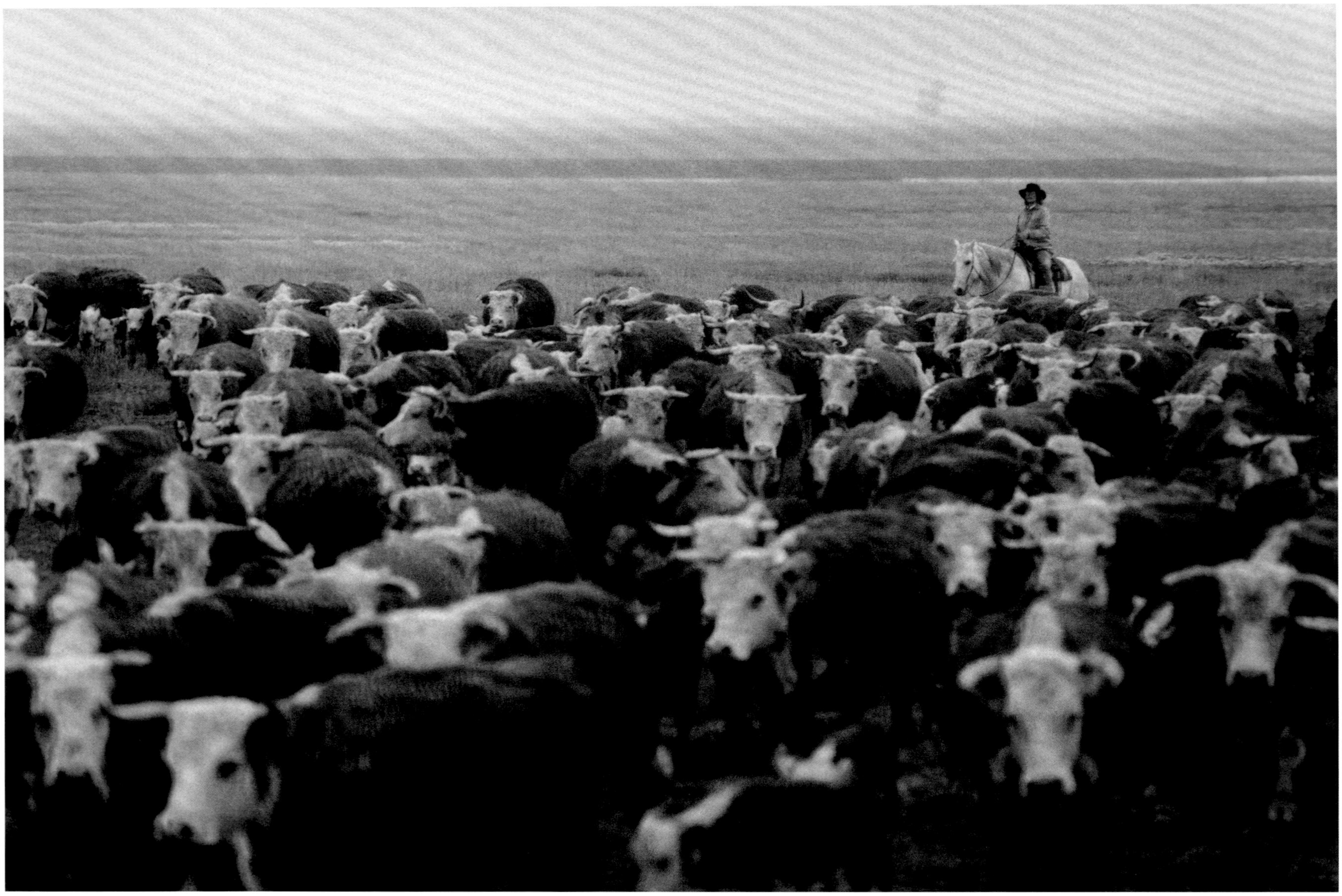

JULIA DAVIS STAFFORD, CS RANCH
1987

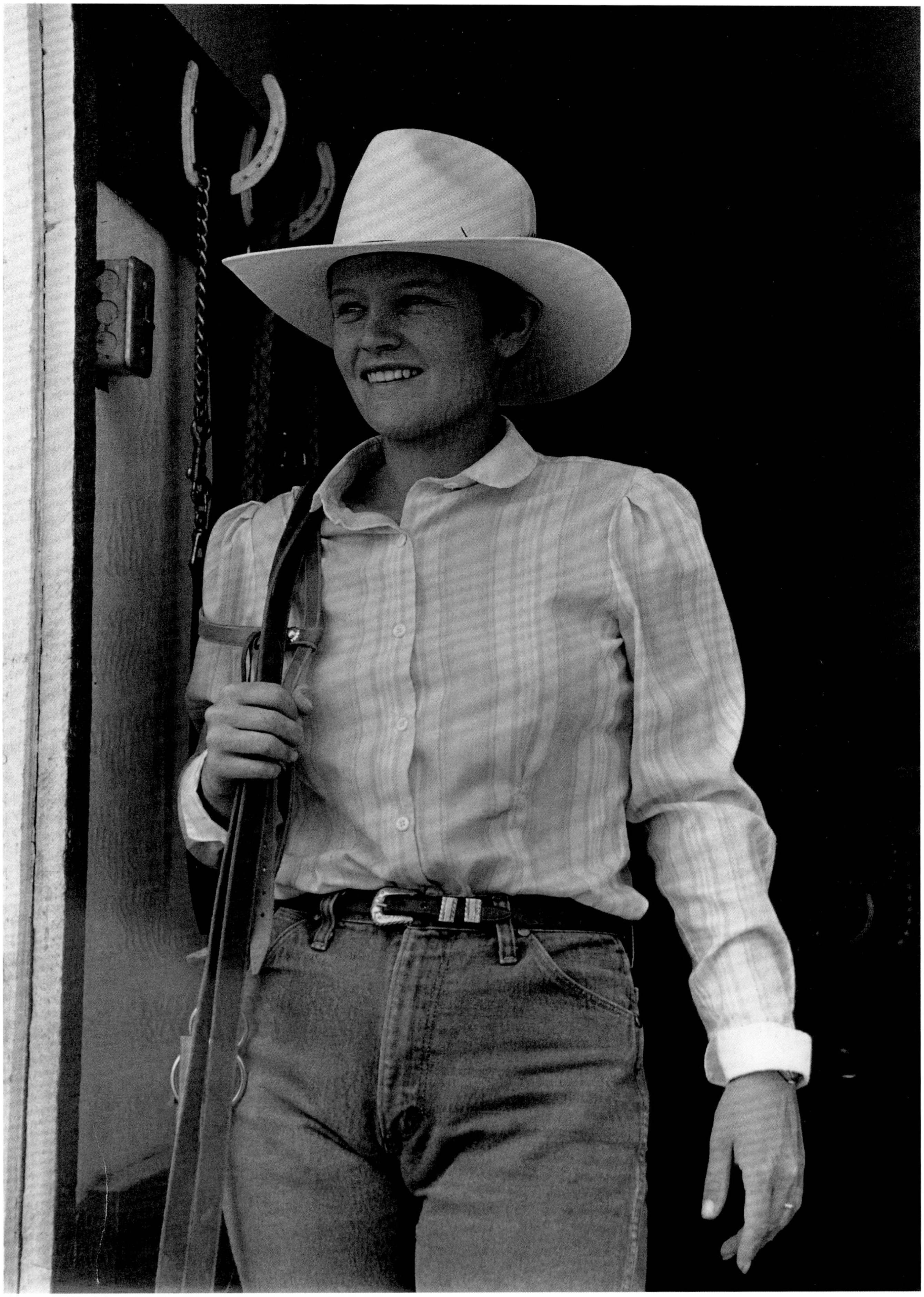

KIM DAVIS BARMANN, CS RANCH
1986

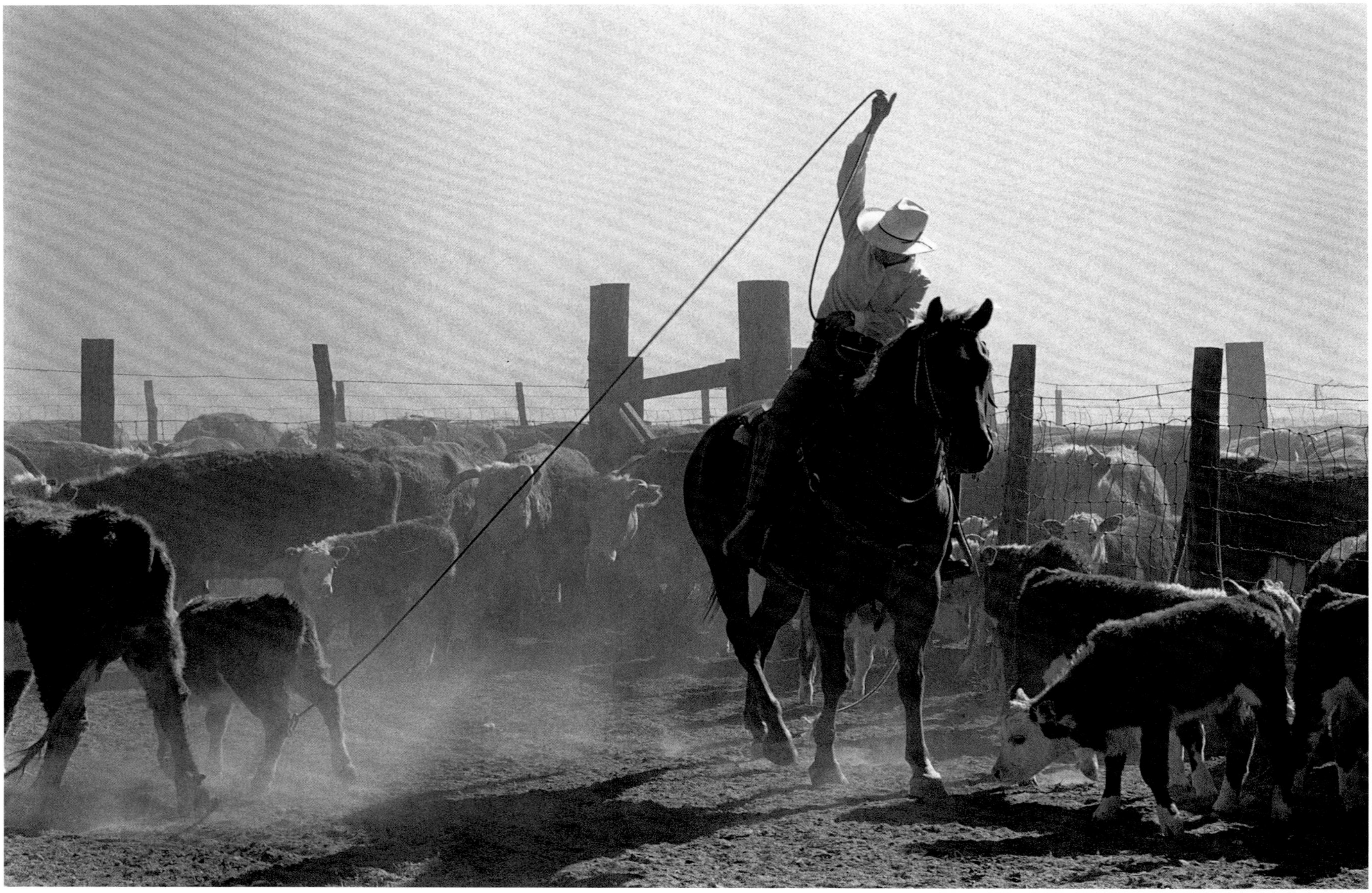

KIM DAVIS BARMANN, HEEL ROPING, CS RANCH
1986

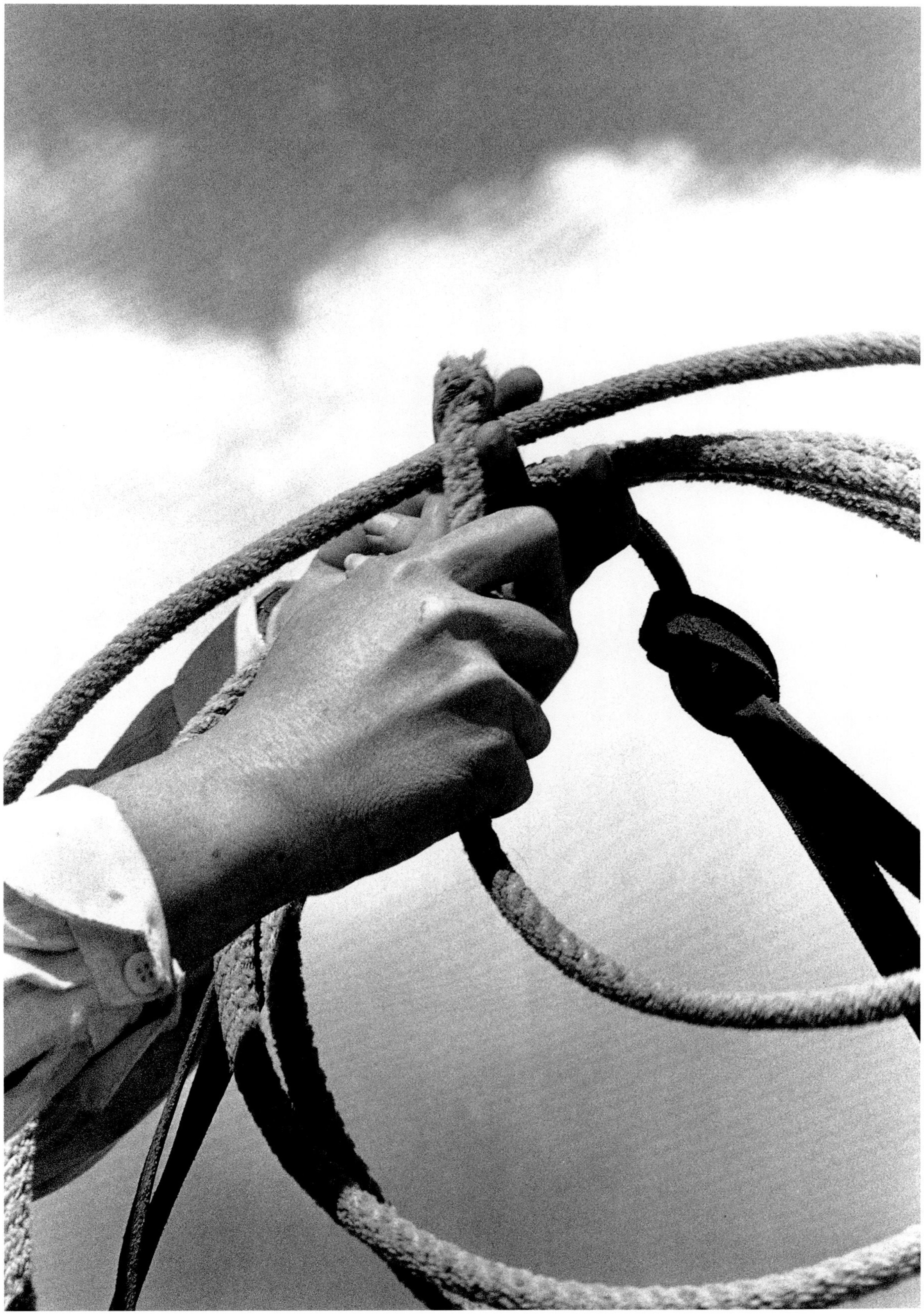

KIM DAVIS BARMANN, "HARD TWIST," CS RANCH
1986

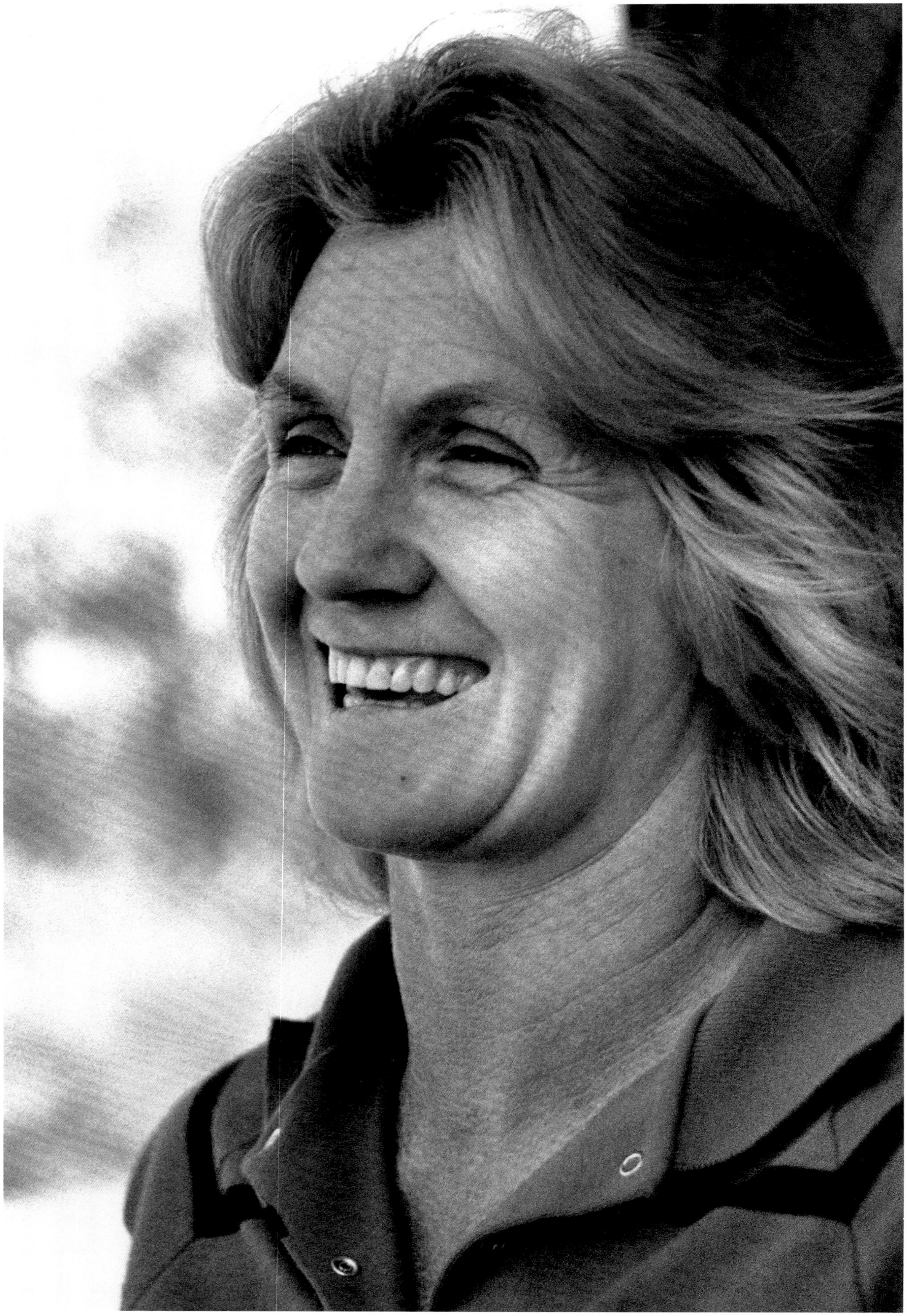

JAN YOUREN, YOUREN RANCH
1986

Jan Youren

Youren Ranch, Garden Valley, Idaho

Jan Youren is a small, wiry, peroxide-blond woman of forty-five who appears to be tougher than nails and nothing to tangle with but is quite the opposite. Jan has a ready smile, loves her eight children and nineteen grandchildren, and although she is a three-time world champion women's bareback bronc rider she is gentle and caring. Like so many women she has had to develop a tough exterior in order to survive. She is solidly married to the man she adores:

Jim is husband number four. Between us we have fifteen kids and then we had one together. I've been rodeoing since I was twelve years old and met my husbands while I was moving around. The first husband was local. The next one I met at a rodeo and it didn't work. Now the third one still rodeos. It was just easier for him not to be married; he couldn't afford to be married and support all those kids! I made a pretty good draw this time; I'll hang onto him! I'd be totally lost without Jim. Sure, I'd bounce back, eventually, but we do everything together. Anytime he's going out he'll say "C'mon Jan, let's go get the cows," or "C'mon Jan, let's go change the water." We get along really swell much as we argue and fight. It's amazing, but good.

Their ranch is located just north of Boise, in the Boise National Forest. This is timber country with grassy valleys and meadows. Jan and Jim's ranch is composed of several different parcels. There is the original homestead place of 300 acres; up the creek is another 200 acres; and then there is another 640 acres of grazing land up in Horse Shoe Bend. They've got a permit to graze 81 head of cattle on forest land. They run about 200 head of cattle in all.

Jan has been riding "rough stock"—the broncs and bulls—for thirty-nine years. "I won $47 riding a bronc when I was twelve years old and thought there would never be another bad day," she laughs. "I'd give anything to be able to ride as good as I did when I was thirteen to sixteen. I could ride 'most anything then. Barebacks and bulls. Generally, Jim stays home and runs the ranch while I'm off rodeoing. Sometimes we get a neighbor to watch it while we go together. He loves that." Why aren't there more women judges? "Not enough women are qualified to be judges and the women are petty, just too petty. Maybe you crossed one of them down the road a few years ago—they don't *forget* it."

Today, many women compete in the women's rodeo association due to the popularity of women riding bulls and bareback broncs. In the old days, they rode saddle broncs; the men, thinking it a kindness, "hobbled," or tied, the stirrups down for the women. That practice caused a number of horrible accidents, and some women were killed. Now the women ride only bulls and bareback broncs, and they have a two-handed rig because most of them are not strong enough in the shoulders and arms to take the jerk. And, because they know the audience and the cowboys don't like to see them get hurt, they will go to great lengths to conceal it if they are hurt.

I saw Jan make a terrific ride on a big, snuffy palomino bareback bronc up in Montana in 1986, but after the pickup man helped her off the horse he accidentally dropped her in front of the chutes, separating her shoulder and spraining her wrist. She rode the next night, all strapped up!

Jan says that it is just part of the risk in all of the fun of rodeoing. One time, after a long drive from Oklahoma to Washington during which she got the flu, she arrived, had no time for a rest, but figured she could at least hold herself together for a six-second bronc ride. She climbed down on her bronc, sick and "zombied out," when ". . . this ol' sonofagun reared over backwards on me in the chute and—*Wham!*—crunched me. Something got messed up in the back of my neck." Then the gate opened and they were out! After the whistle blew, she let go and "landed like a sack of 'taters'—hard—and popped a rib." Never mind, she rode the next day too.

We don't say we're as good as the men, but that we have a place of our own. The guys have been tremendous in helping us at rodeos. They'll help you "pull your rig" and give all the advice. Anything they can do to help—they're really swell. But it didn't start out that way. Back at the Cow Palace in 1979 the cowboys were not for it—they were against it. Bruce Ford said, "I thought these gals were all out here trying to act like guys and be really macho." But he sure changed his mind and admits it. Tanya [her oldest daughter] won it the first year they had the women's rodeo at the Cow Palace. She marked a ninety-one on one of her bronc rides and sure impressed ol' Bruce. The judges are all men, except in our finals, where there has to be at least one woman judge.

Somehow, between all the years in the rodeo Jan managed eight healthy pregnancies:

I rodeoed up to when I was five and a half months pregnant. I don't have any trouble having kids. I went to the hospital for the first five; the last two I just went to the doctor's office, had 'em and came home. And the last one I had at home. One daughter was born about 8:00 in the morning and that night I went to a dance we were having. But I made a mistake when Susie was born because I went back to riding broncs before she was two weeks old. . . . I started to bleed pretty good, but I didn't tell anyone. I knew I shouldn't be riding after the first bronc, but I rode the second one. Then I didn't have another rodeo for two weeks, and by that time I was in good shape.

The rodeo life has been good to Jan and her kids. They've traveled all over doing noncompetitive rodeo exhibitions:

I think it's been a good life. It's been a pretty good one for the kids. They haven't been into dope or anything like that, but then I threatened them with their life if they did mess with it. I don't even want them thinking about the stuff. Tanya, my oldest girl, got sick after a tequila drinking contest. She came home sick, sick, sick. When I came home and found her lying down, I said "What are you doing? You've got chores. There's work to be done!" My husband said, "Oh, Jan, you're such a meanie. She's so sick. I know how she feels." But I said, "Well, she's going to remember this for a long time. I'm not gonna let her just lay there wallowing in her misery. She's going to get up and do her work!" [Laughs] I didn't have any more trouble with her.

Western Horseman magazine March 1995:

"Fifty-one-year-old Jan Youren became the oldest woman to win a world bareback bronc riding title. Youren, a mother of eight and grandmother of nineteen, clinched her third world bareback riding title with a dramatic ride during the final day of the November 4–6 Women's National Finals Rodeo at Fort Worth's Cowtown Coliseum." In 1993 Jan Youren was inducted into the National Cowgirl Hall of Fame.

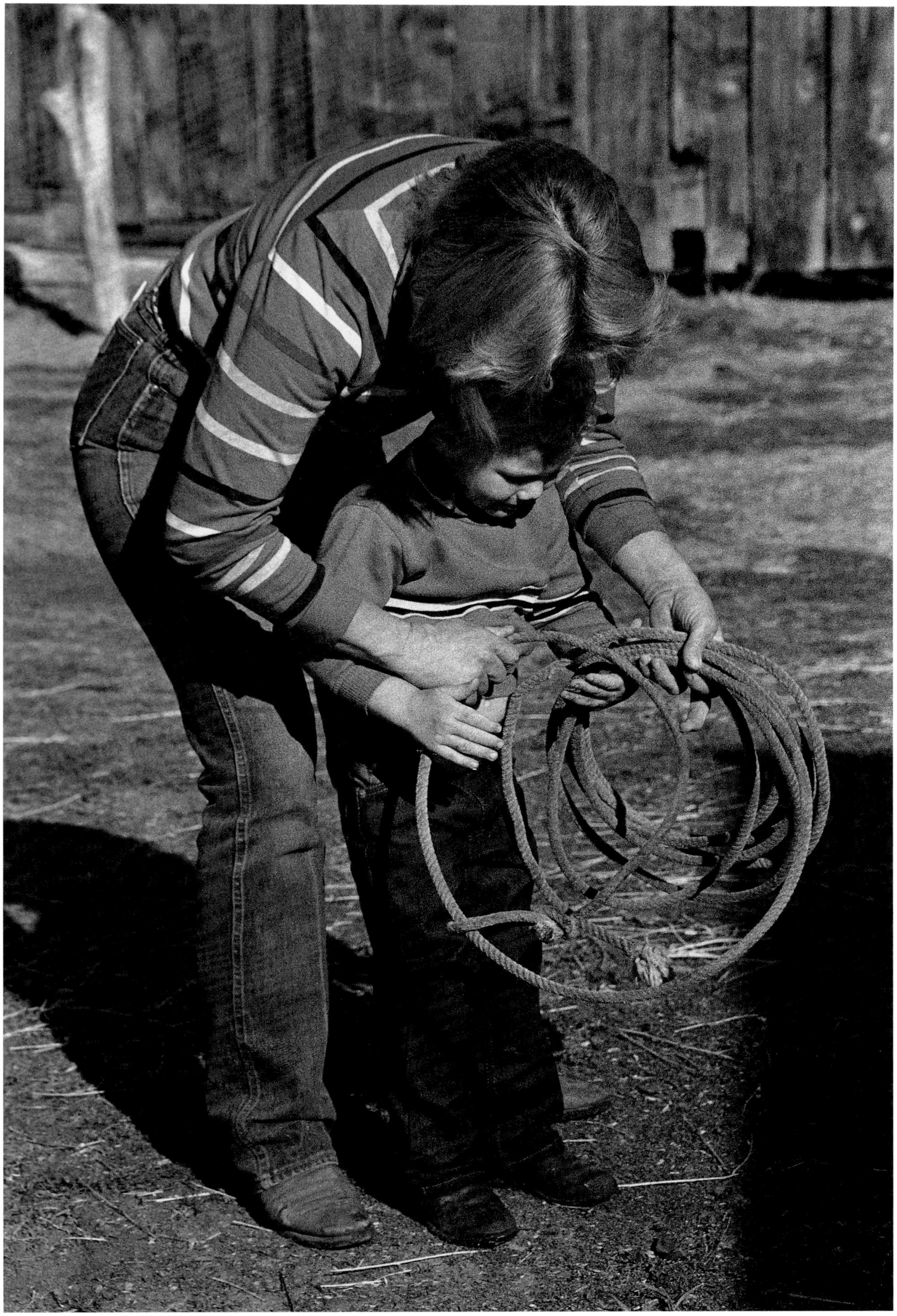

JAN YOUREN AND GRANDSON, YOUREN RANCH
1986

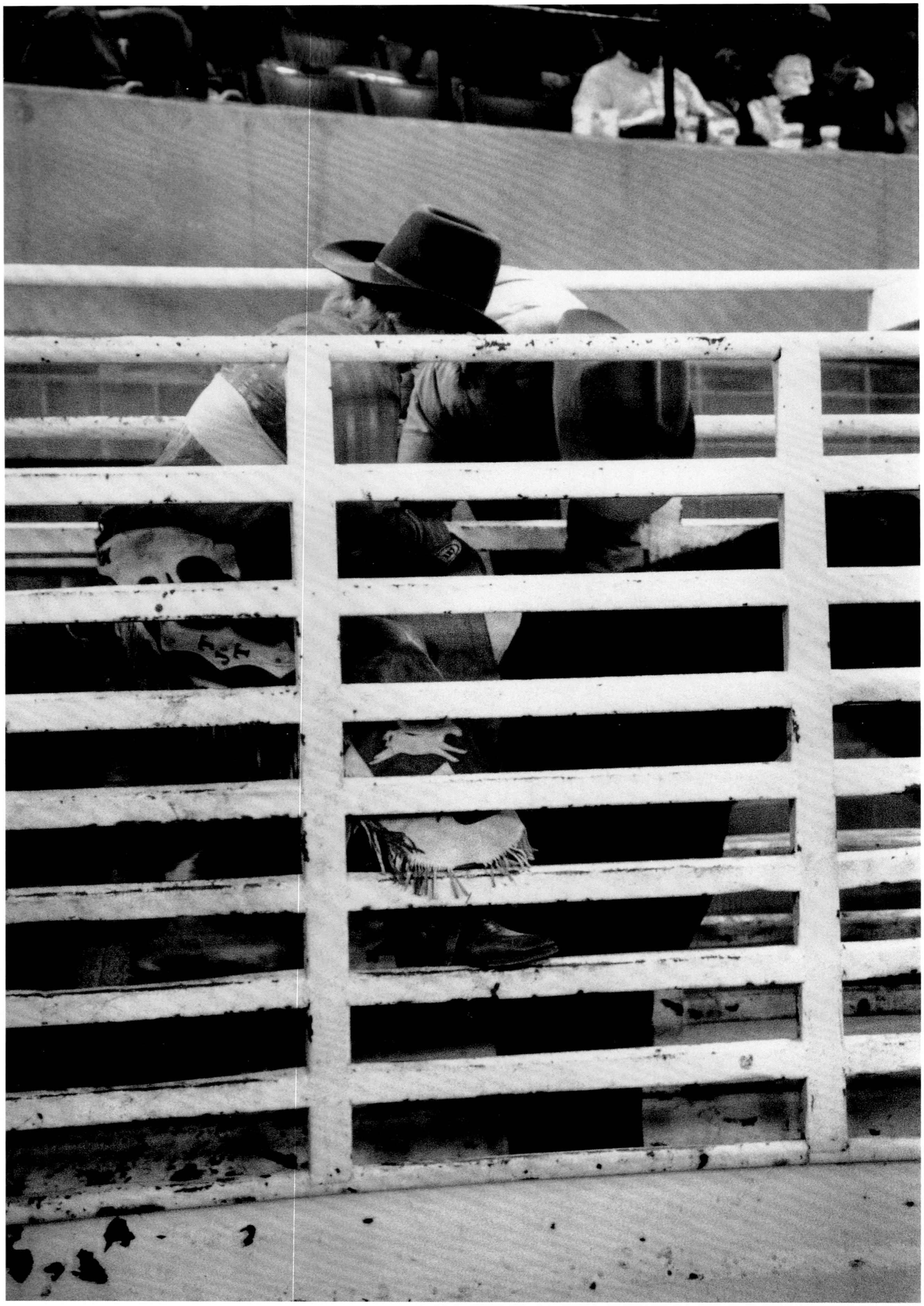

JAN YOUREN AND HUSBAND, JIM, AT BILLINGS, MONTANA, RODEO
1986

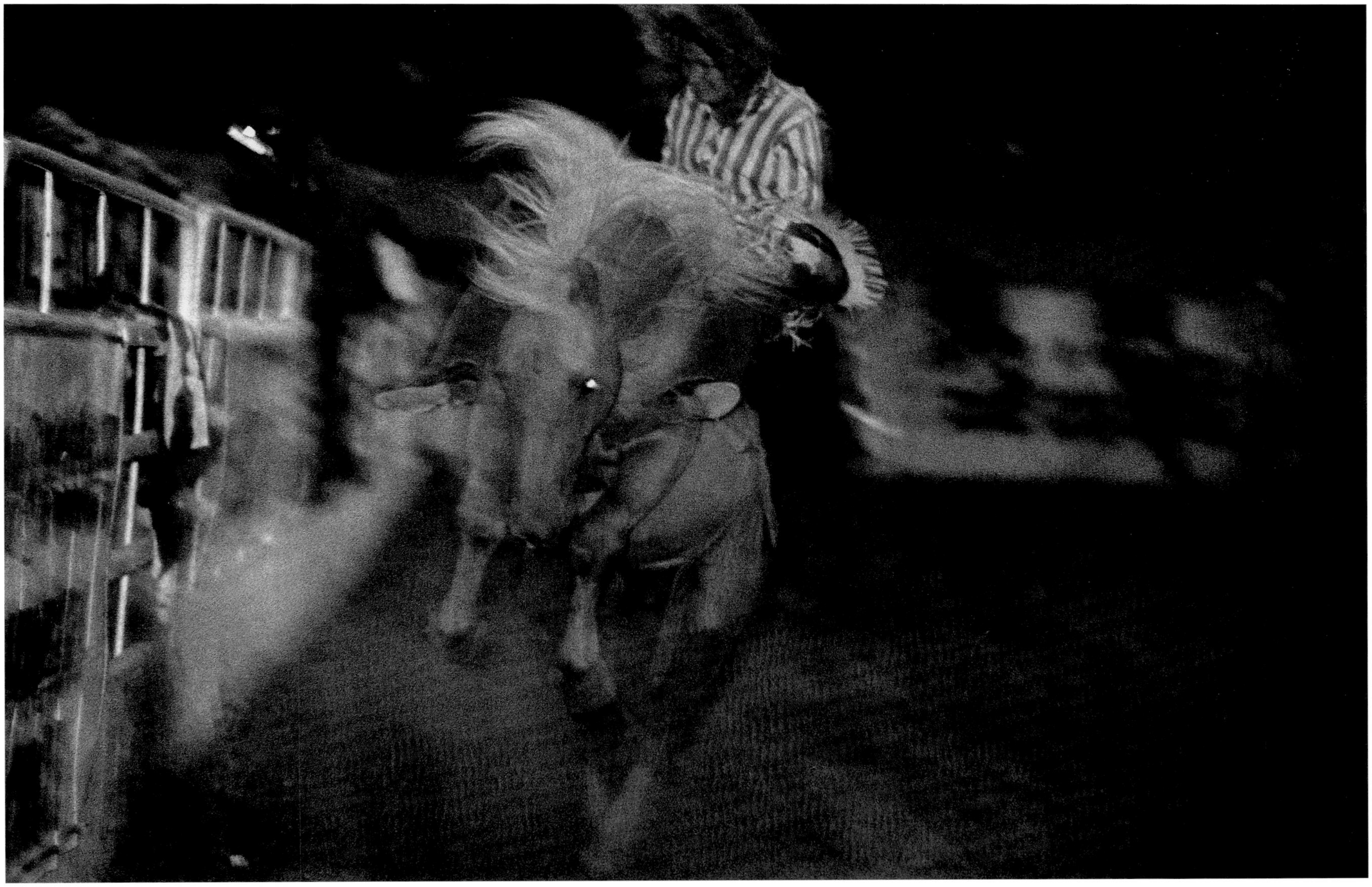

JAN YOUREN, BILLINGS RODEO
1986

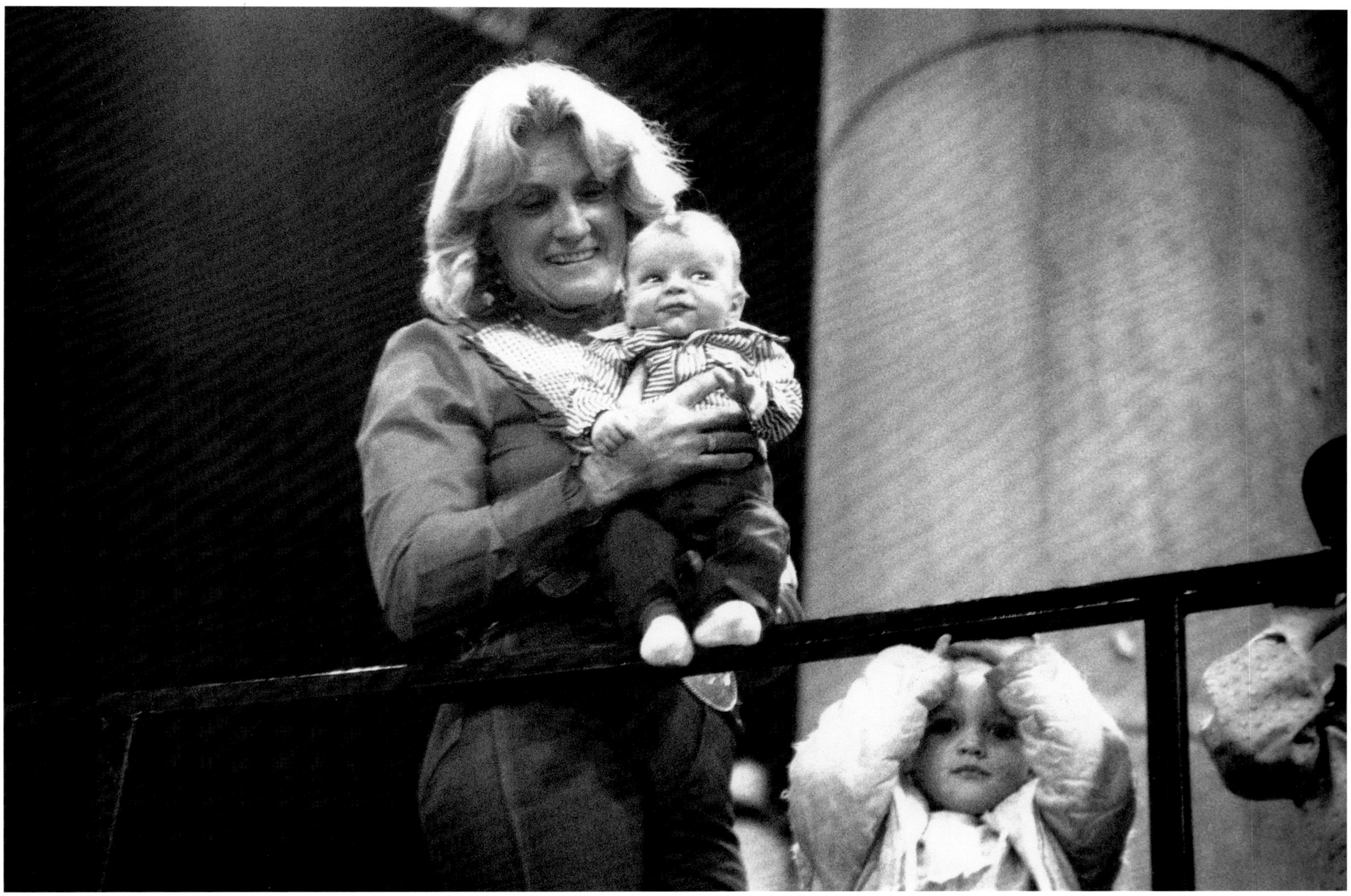

JAN YOUREN AND GRANDCHILDREN, BILLINGS RODEO
1986

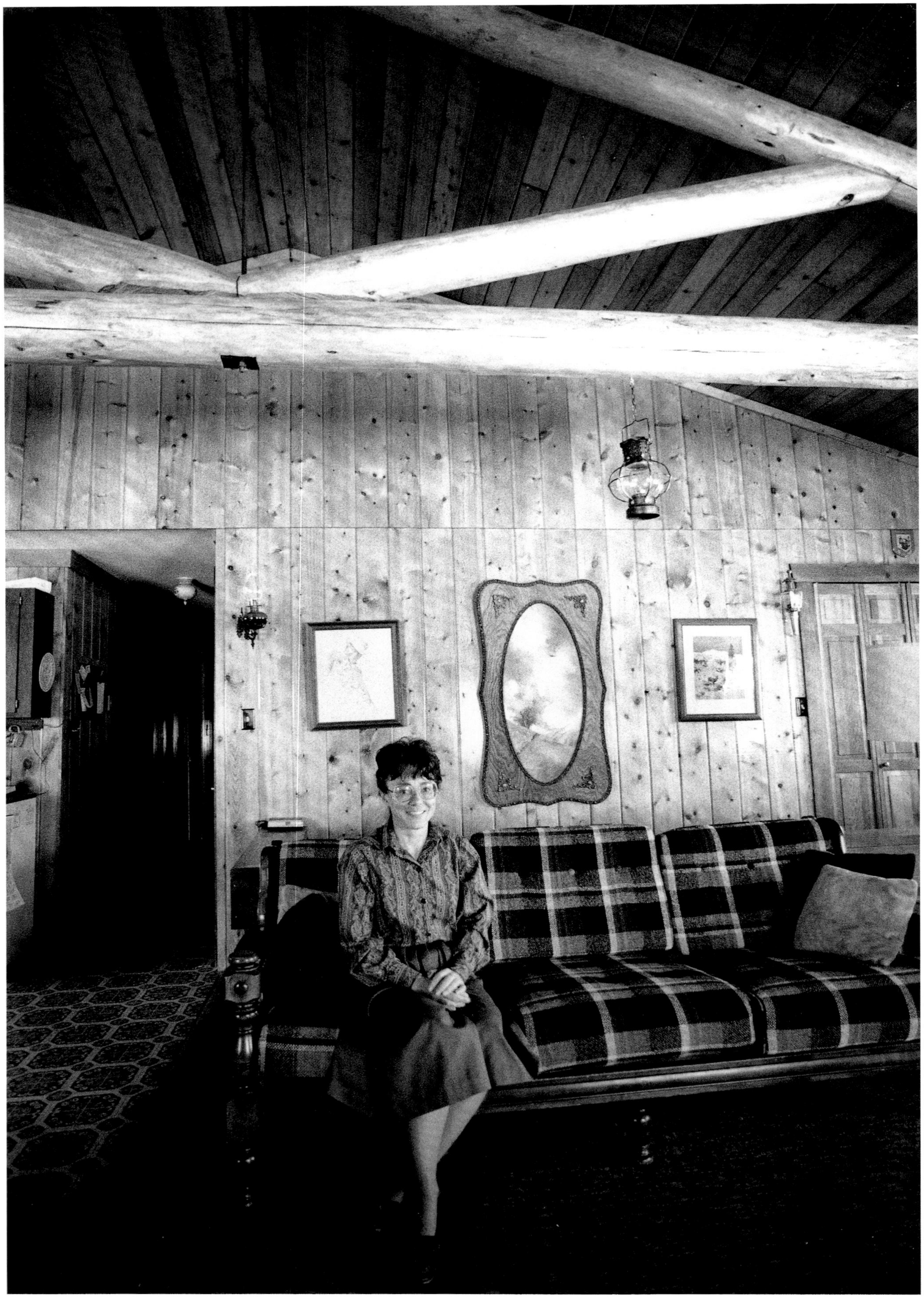

MICHELE CARROCCIA, OTTER CREEK RANCH
1988

Michele Carroccia

Otter Creek Ranch, Melville, Montana

A small, raven-haired, freckle-faced woman of fifty-four, Shelly, as she is usually called, lives with her husband and two of the four children they have raised on the east slope of the Crazy Mountains in south-central Montana. Their summer place, which they run as a dude ranch during the short Montana summer, is located in Sweet Grass Canyon, a wide U-shaped valley through which runs the creek by the same name. Their winter ranch, the Otter Creek Ranch, is located about ten miles to the southeast where the rolling prairie meets the foothills. It is dramatic country and one that has been good to her family.

Her ancestors have been in Montana since 1836 and have been ranching in the same area since 1880—a matter of family pride. A former school teacher, she is articulate and, like most ranch women, outspoken:

I've aired my views on the harassment of individuals by the federal government. People need to accord farmers and ranchers respect for the work they do. We're not looking to be patted on the back but neither are we enjoying being looked down on as "hicks." If we weren't businessmen, we couldn't stay in business. We just go about it in a different way. We still value a handshake and knowing that a man's word is good. Paper burns. Words don't.

Shelly was born and raised on the winter ranch that she, her husband, and her brother now own. She married Bill, another native Montanan, just after college, and they went to southeastern Alaska as teachers. On a trip back home during the summer, she showed Bill the Sweet Grass Canyon and the ranch her grandfather offered to sell them if they would move back and run it. Bill fell madly in love with the spectacular beauty of it, but Shelly saw it from a different perspective. They bought it and moved in. While Bill taught school in town and came out during the weekends, she stayed at the ranch with their two babies. One cold December the pipes froze and there was no water. "It's not my idea of fun to haul water from the creek for two babies in diapers."

Shelly is non-stop motion. She had always sewed shirts for her family, so soon enough she started sewing the old-time, full-length dusters for her family and various guests from their dude ranch. Then she moved into quick-freezing and one-day shipping the prime cuts of the grass-fed, hormone-free beef they raise on the ranch. "The overhead has gone up some two hundred plus percent since 1955 and the prices for cattle are about the same now as then. So no, you can't support a family by ranching alone. That's why people have to diversify."

Asked what she likes best about ranch life, she continues:

I like being my own boss. I like being responsible for the grazing, for taking care of the land, for being good stewards of the land. I like working with the livestock, and I love looking at the mountains. I resent the federal government, the bureaucracy with no hands-on knowledge of the land, coming in and telling us "You have to be good stewards of the land." If we didn't take good care of it, we would lose it.

She and Bill work together on all aspects of the ranch: haying, moving cattle, shipping, fencing, as well as the dude ranch operation. She helps Bill, Tack, her brother, and Tony, her youngest son, tend to the long hours of calving. All the children grew up helping on the ranch, but now only the two youngest sons are active on the ranch. Their only daughter is married and a lawyer, practicing in the state. Shelly is an unabashed champion of the superiority of raising children on a ranch:

Ranching is the best possible way to live—a great way for kids to grow up learning responsibility. They learn to work together as a unit. It does away with the so-called generation gap because you do everything together. Kids know that they are necessary; they're respected; they gain a strong sense of self worth and family unity.

True to form, she is conservative and has strong religious beliefs and practices that permeate her daily life.

I'd rather be outside riding, working cattle than any place else in the world. It's a feeling of belonging; everything is right. When I'm out in God's country, everything fits into place perfectly. There's a feeling that this is the way that the world and life should be. You've got to work; you've got to put up with nature and the elements, but it's right. It just fits.

Warm, gracious, and community-minded, she even served as justice of the peace for a number of years while she was teaching. She embodies the West's spirit of independence:

I do not like passive entertainment. I like to be actively and mentally involved. If women ranchers always had to have company around, we wouldn't have a West. There is no lonely. You don't rely on someone else or something else to entertain you. Everything has to come from within yourself.

I think rural women are capable of efficiently dealing with almost any emergency that arises. If the cows are in the haystack, if a calf is born in the creek and has to be brought in and put in the bathtub . . . , survival of either your livestock or yourself takes precedence over everything else that can happen. Sure, you kill [wild game for food] and you butcher some of your cattle too. That's why they are here. We enjoy seeing them, but we also use them as food, and we use their hides. We never take for granted that life, and we never waste it.

She is equally as forthright about her marriage.

We ranch women tend to bypass the need for the frills of romance and just get down to the meat of the matter in marriage, which is companionship. My idea of a perfect anniversary is to go to the head of South Fork, just Bill and myself, and camp out, cook over an open fire. There's nobody there but us and the mountains and the animals. You can't miss the fact that God has a hand in all this, and that really is what the bottom line is—that it's Bill, me, and God. That's the bottom line in marriage too.

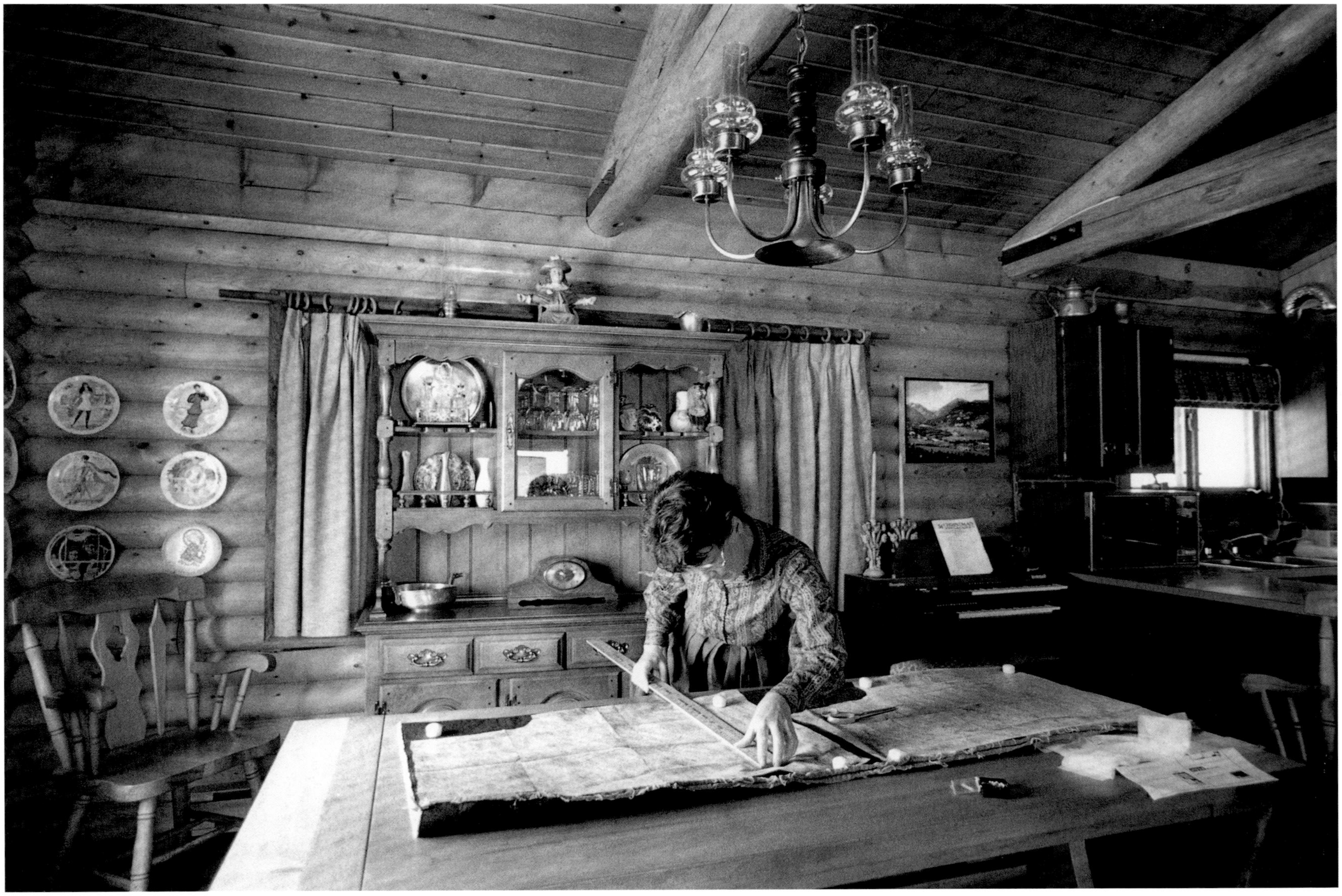

MICHELE CARROCCIA, OTTER CREEK RANCH
1988

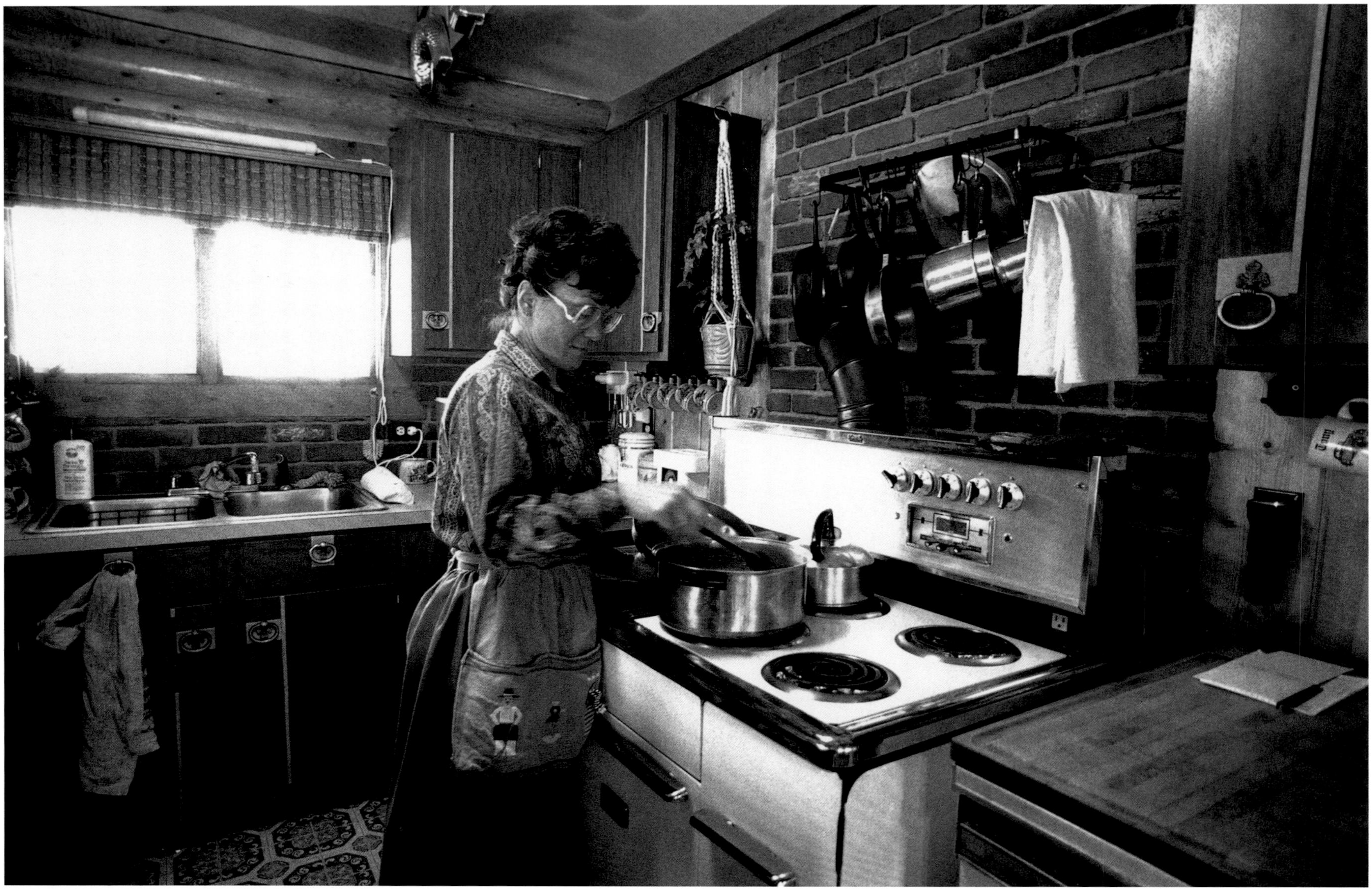

MICHELE CARROCCIA, OTTER CREEK RANCH
1988

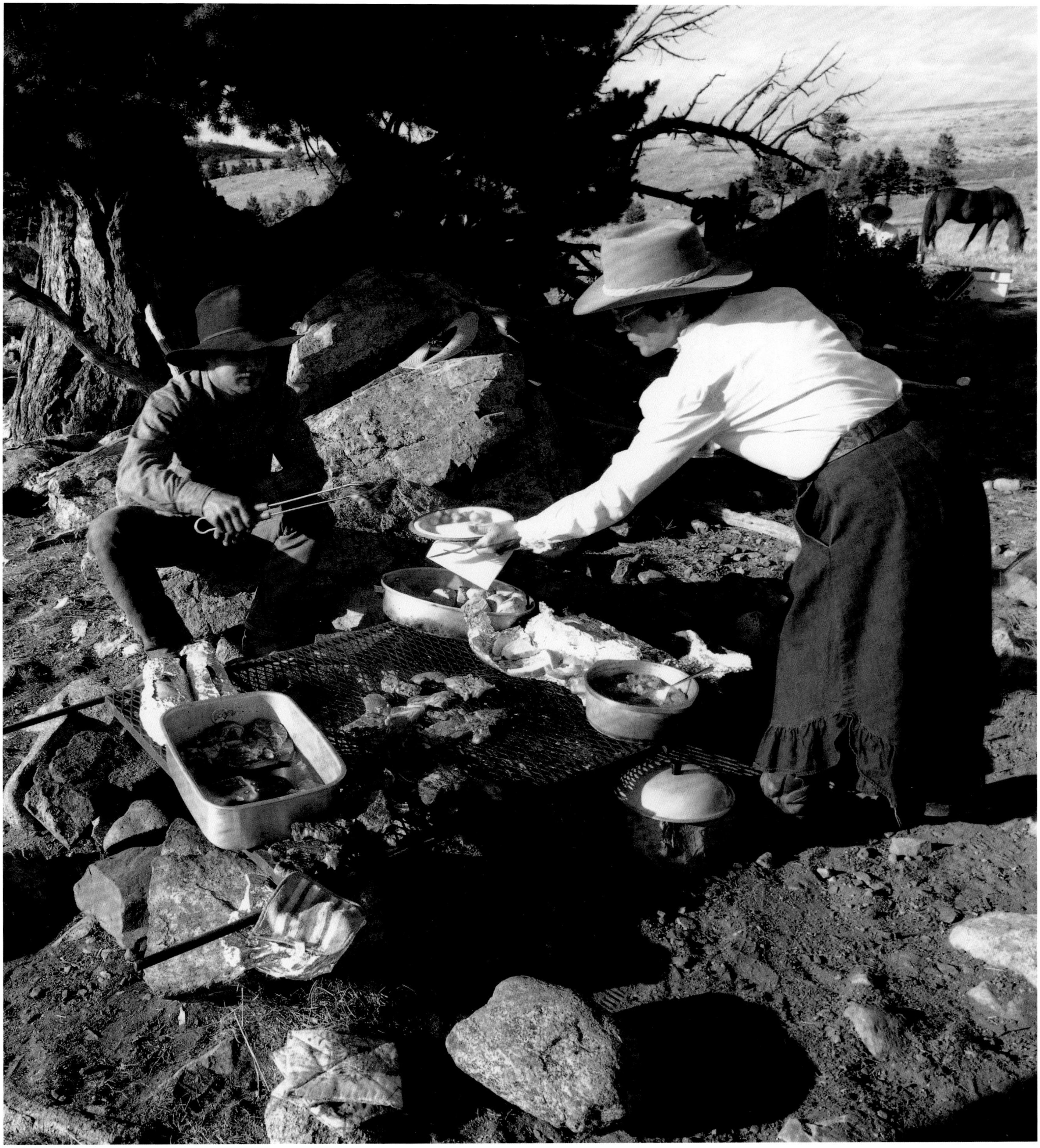

MICHELE CARROCCIA AND SON ROCCO, SWEET GRASS RANCH
1988

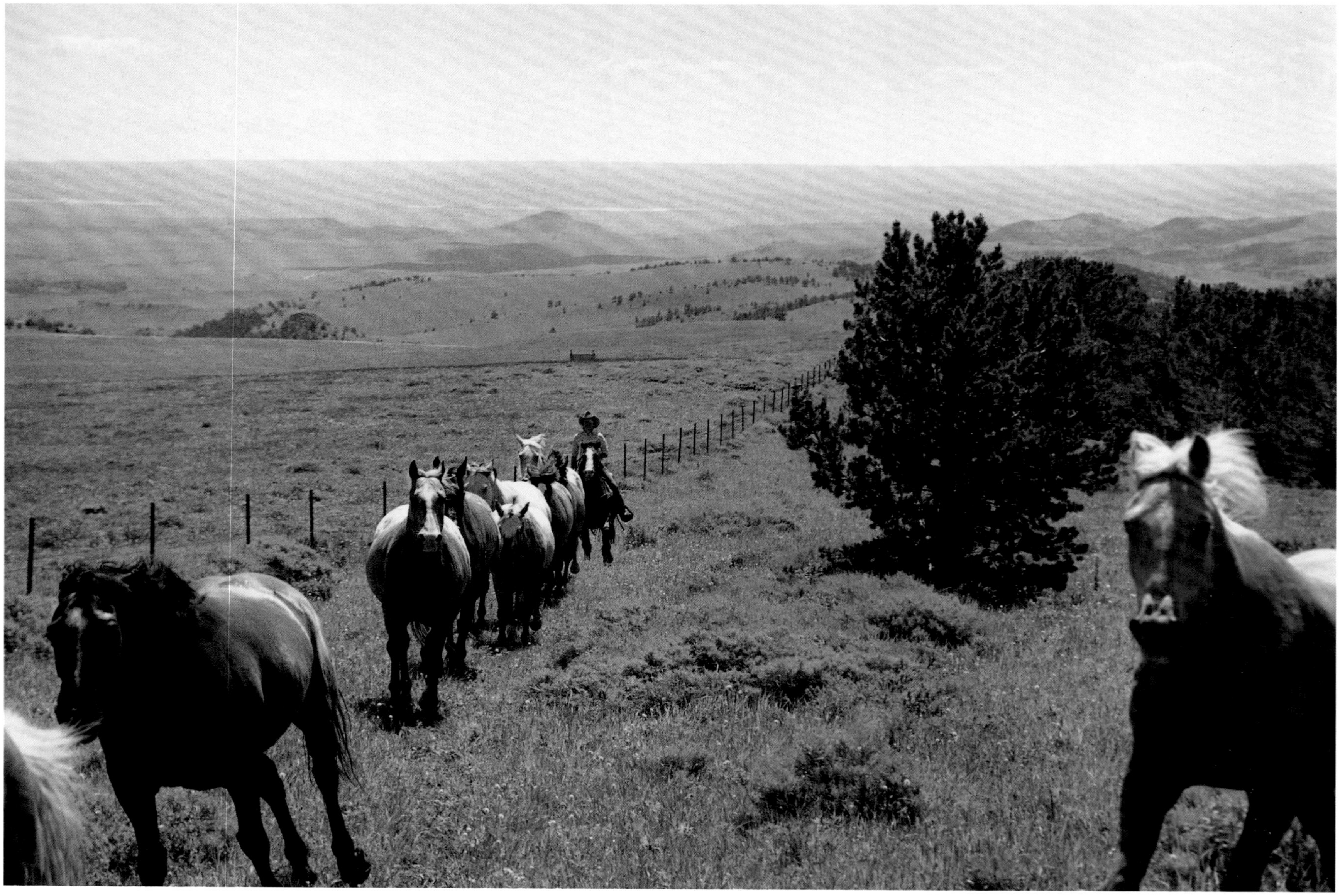

MOVING HORSES, SWEET GRASS RANCH
1986

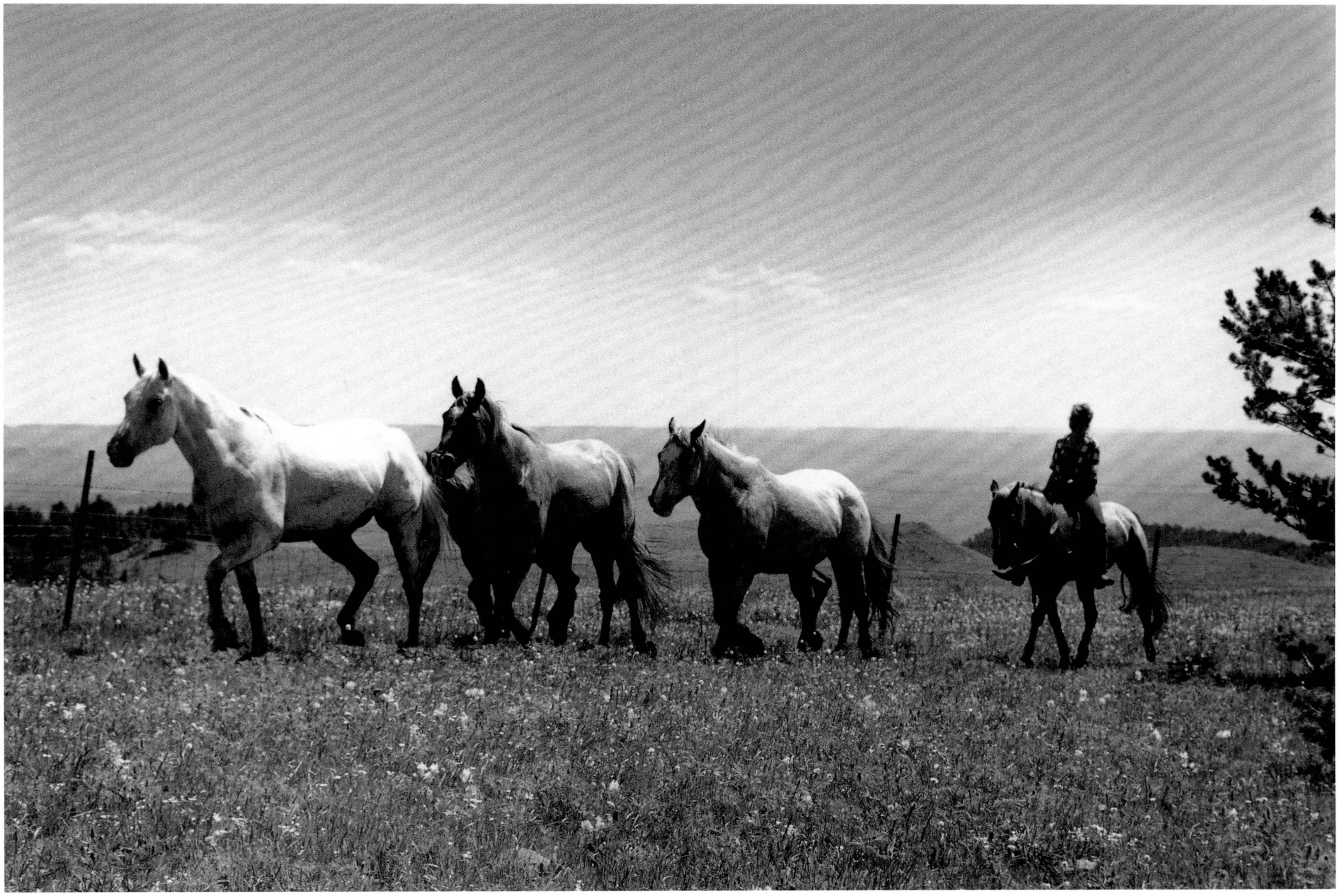

PAGE DRINGMAN, SWEET GRASS RANCH
1986

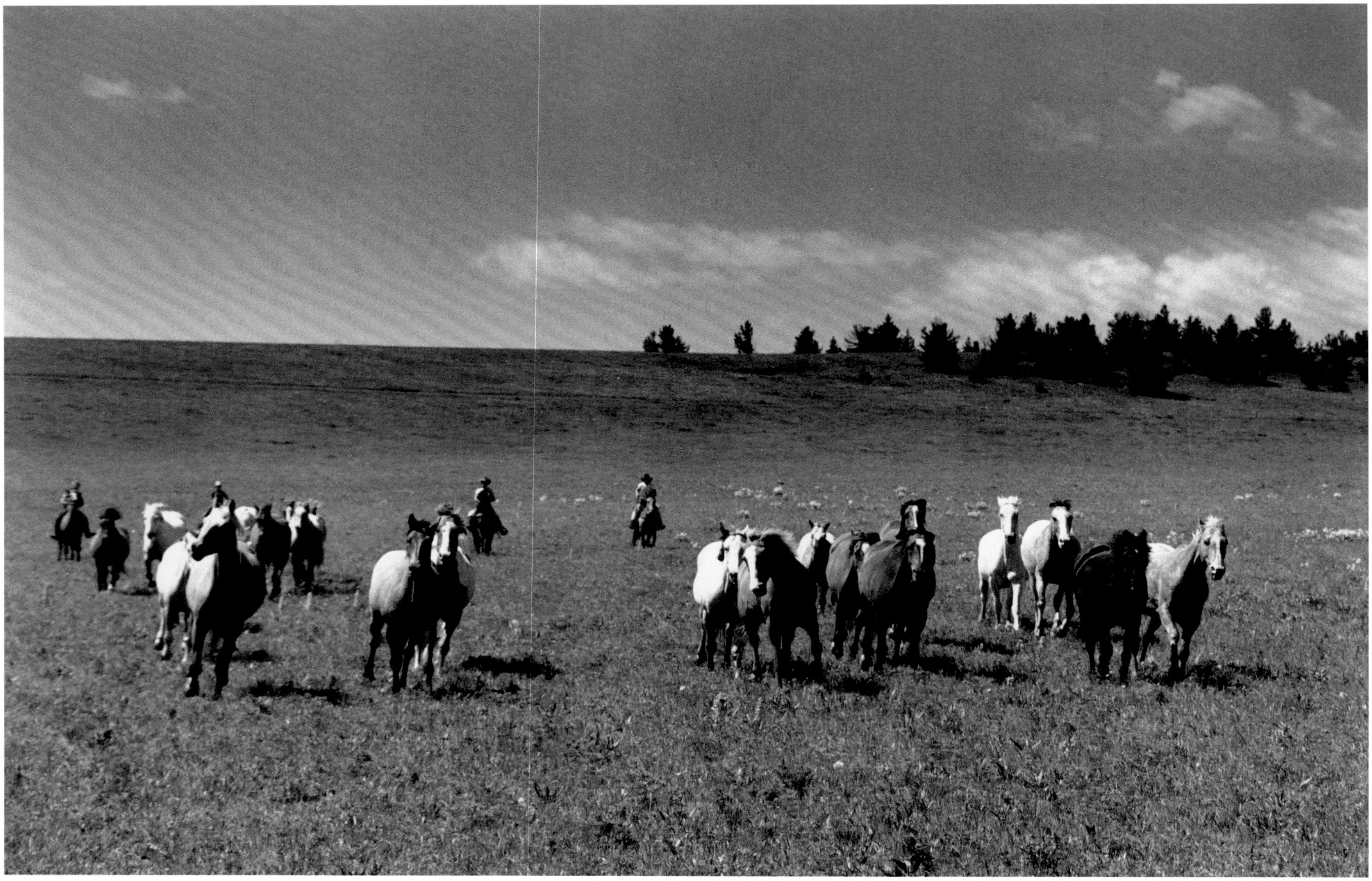

CARROCCIA FAMILY, MOVING HORSES, SWEET GRASS RANCH
1986

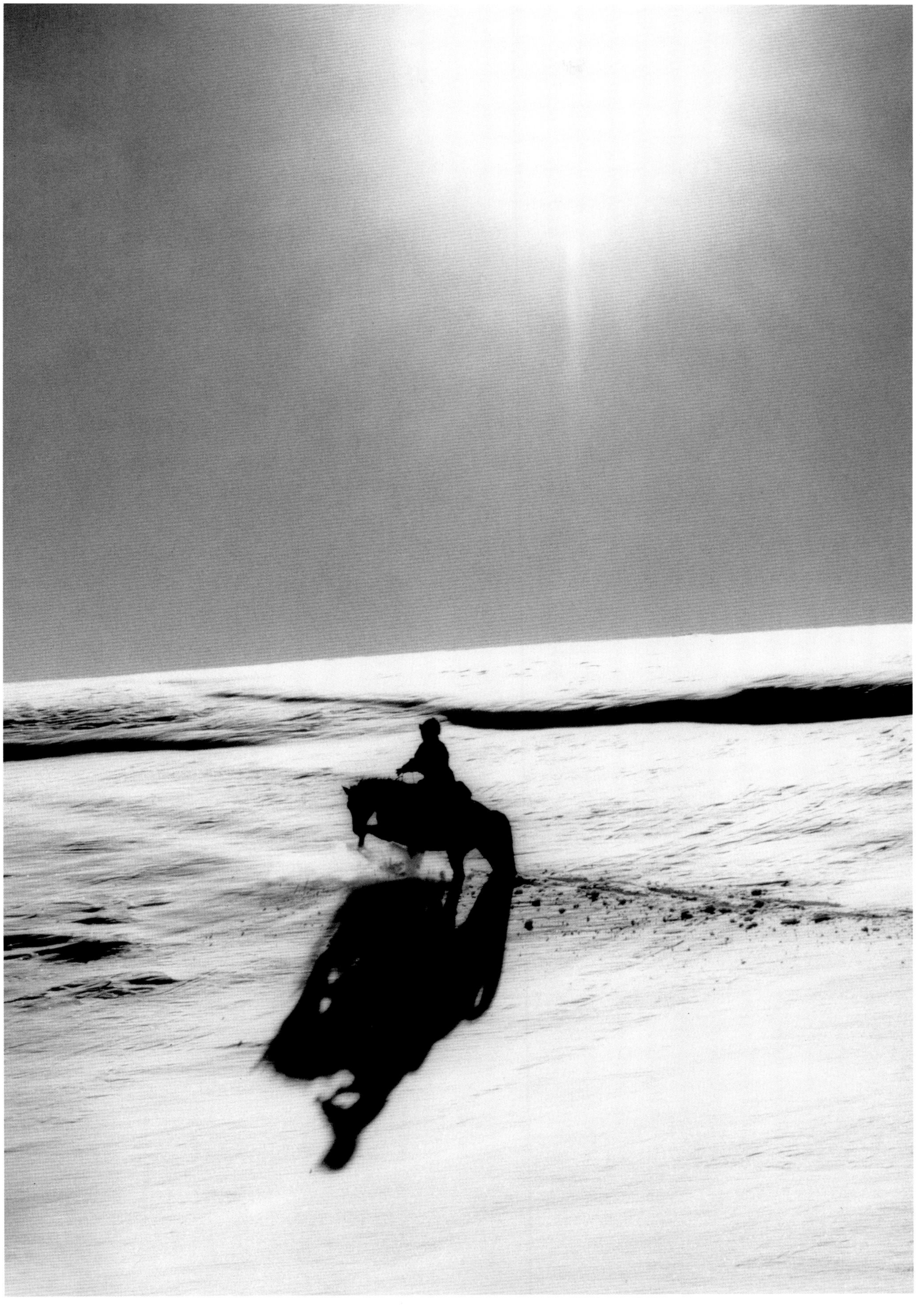

MICHELE CARROCCIA, OTTER CREEK RANCH
1988

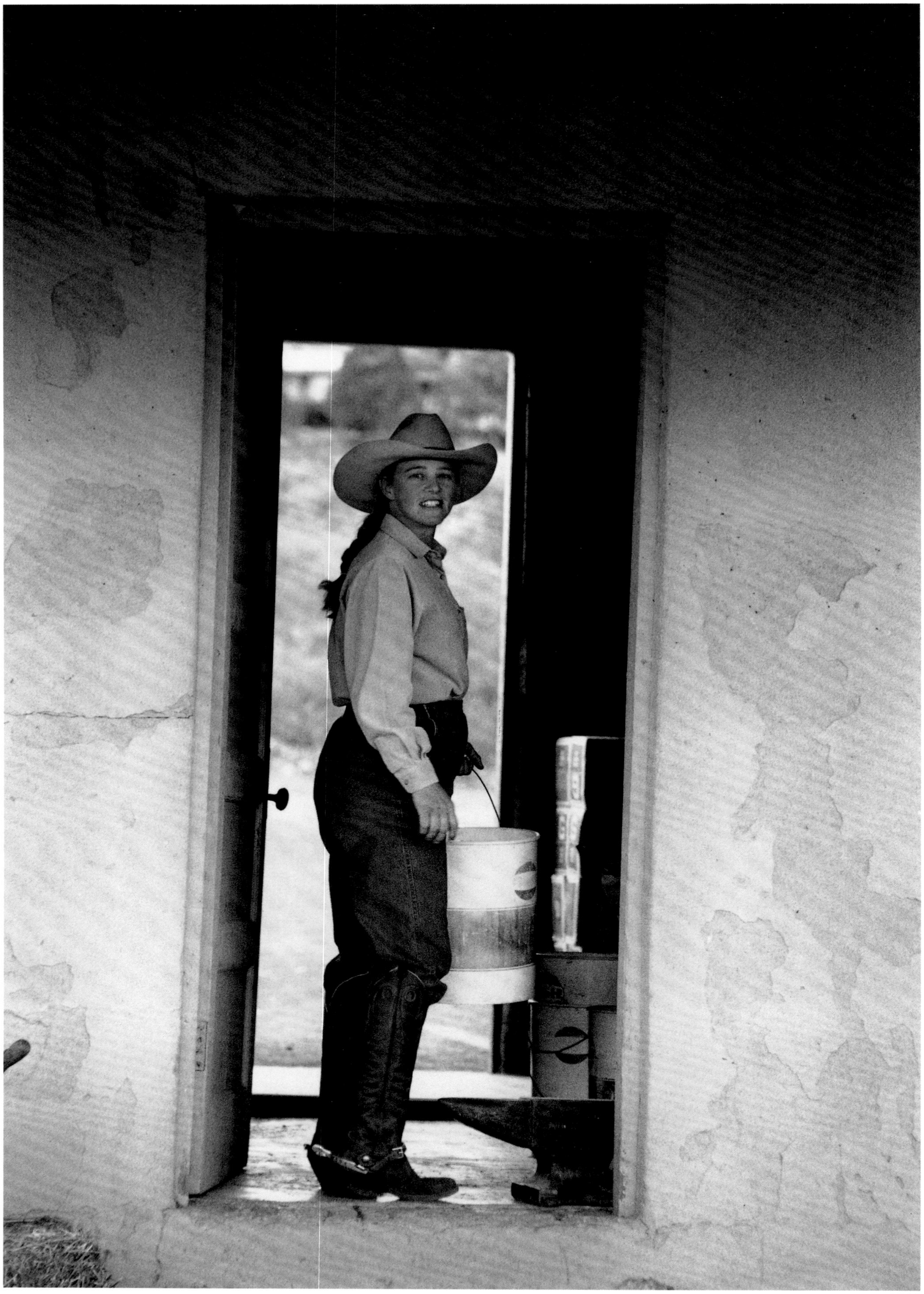

ANN HOLLAND DAUGHERTY, GAGE HOLLAND RANCH
1988

Ann Holland Daugherty

Gage Holland Ranch, Alpine, Texas

"Be at the ranch at 4:30 a.m." Ann said, and then gave directions that included lots of "forks in the road," "bear left" or "right," and "don't take the faint tracks." Following ranch roads and country directions can be challenging enough in broad daylight, but at 3:30 in the morning all I could see was black. I regretted not accepting Ann's offer to stay at the ranch. The night dew was heavy and it was almost freezing. The windshield on my pickup frosted over, making it hard to see the tracks and follow the landmarks of Ann's directions. But, there were two pickups ahead seeming to head in the same direction. I hoped the drivers were ranch hand day-helpers headed to Gage Holland Ranch, and they were.

We all arrived and were invited in to breakfast at the cook house. Later, under a canopy of glittering stars that seemed to be within arm's reach, I saw the faint tinge of purplish pink on the eastern horizon. Soon the hard night air would soften and warm. The horses were saddled and loaded in the stock trailers to be hauled some miles closer to the cattle to be gathered.

Ann is twenty-seven years old and four months pregnant. She has been married to Mark for five years. A lovely, blonde woman, Ann was severely bitten on the right side of the head as a child by a dog that belonged to her uncle. As a result she is "high-tone" deaf. "I can hear a deep voice or low sound pretty well but I can't understand a woman's or child's voice until I get used to it, then I can guess at what is being said."

After college course work at Texas A & M and Sul Ross College, she finished at Texas Tech with a degree in animal production. "There never was a doubt in my mind that I would go to college, major in some kind of agriculture, animal production or range management, graduate, and then come back fast as I could and run the ranch. I wanted to run Daddy's registered herd."

Originally when her grandfather owned the ranch it was a big sprawling outfit about which Ann says, "I heard a person could travel from the San Francisco area to Fort Davis on land that was owned or leased by [Grandfather] Gage." Today the West Texas ranch north of Big Bend encompasses a rolling, broken-hills grass country with lots of rimrock sides. It must be good country because the cattle, Braford cross (Hereford Brahma), looked good.

Ann and Mark met at college and returned together to the family ranch. "After we got married, I had to grow up a lot. I had to put Mark first instead of the ranch being first. That's hard to do, but if I didn't the marriage wouldn't work. If the ranch comes first, then Daddy comes first, and that's not what marriage is about."

In the early years of their marriage, Ann and Mark worked alongside Ann's father and tried to benefit from his lifelong ranching experience. "We both want to know how Daddy does something, why he does it, and what happened when he did it wrong. We both figure that Daddy has done it every way including every possible wrong way. So we figure that if we can learn from him and do things right the first time we won't waste time and money."

"It is important for me to be here on the ranch. If we weren't here, Daddy wouldn't have any of his family here, and all of his work to keep the ranch going would seem like wasted effort. My being here helps to keep his interest in the ranch alive. He can think of grandchildren and of the ranch continuing in the family and not having to be sold. What purpose in life is there if you don't have someone to pass on the fruits of your life's work to?"

She worries about the business end of the ranch. "I can do the physical work and I know about the livestock management, but as far as making sure we have the right kind of insurance or what kind of taxes to pay or which things are capital gain or income tax and so forth, I do not know all of that. Yet."

I ask what she would do if Mark died. "I'd go on alone running the ranch. I think men would probably work for a widow better than they would for a woman who has never been married. On the ranch, I find that the men tend to want to take care of me. That is great and I love it. But sometimes they act as if I can't do my job, which is most definitely not the way to act! A lot of times I tell the men to do something and they don't take me seriously which makes me fighting mad. I've been on this ranch a lot longer than they have and I *do know* what I'm doing."

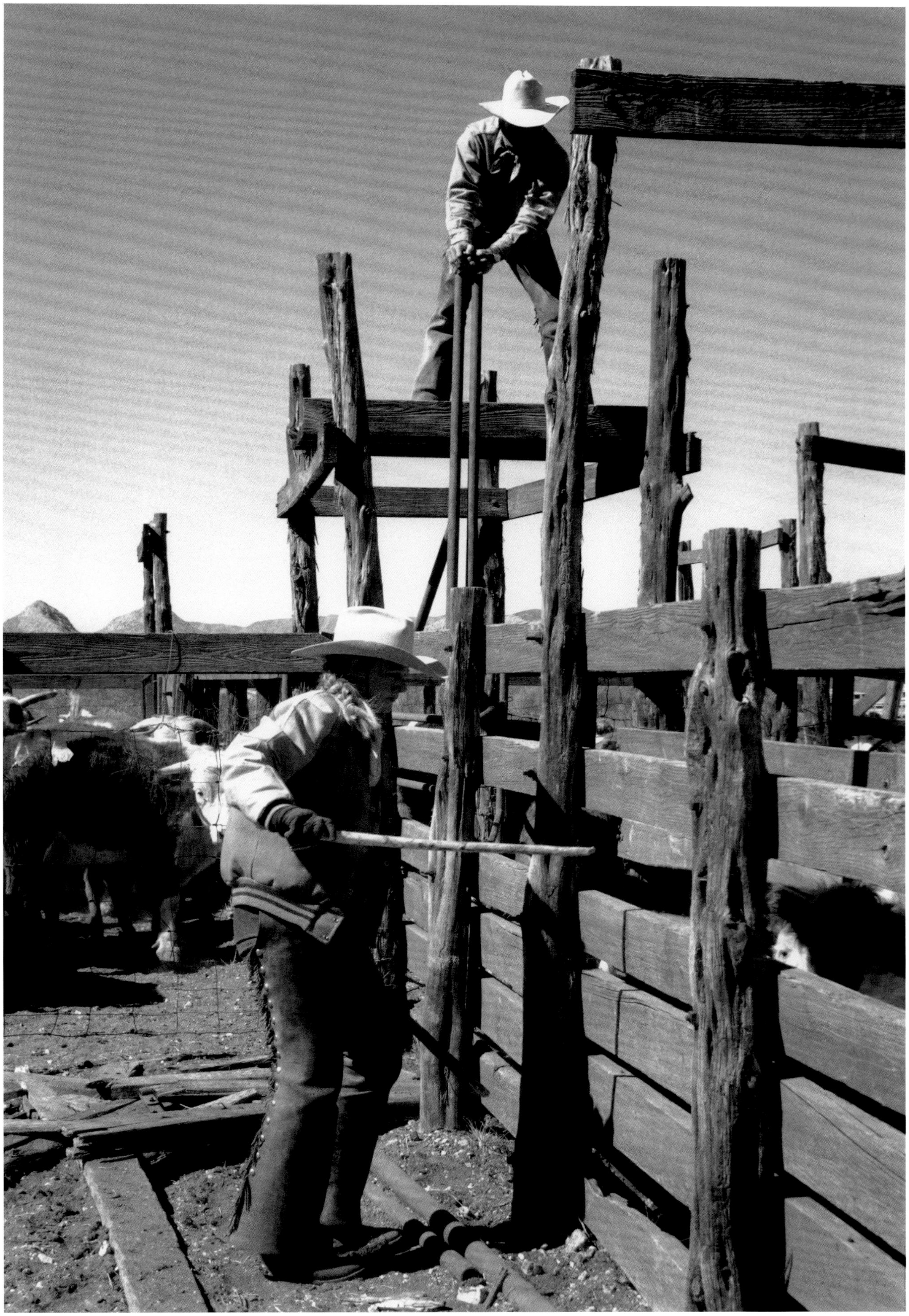

ANN HOLLAND DAUGHERTY AND HUSBAND, MARK, GAGE HOLLAND RANCH
1988

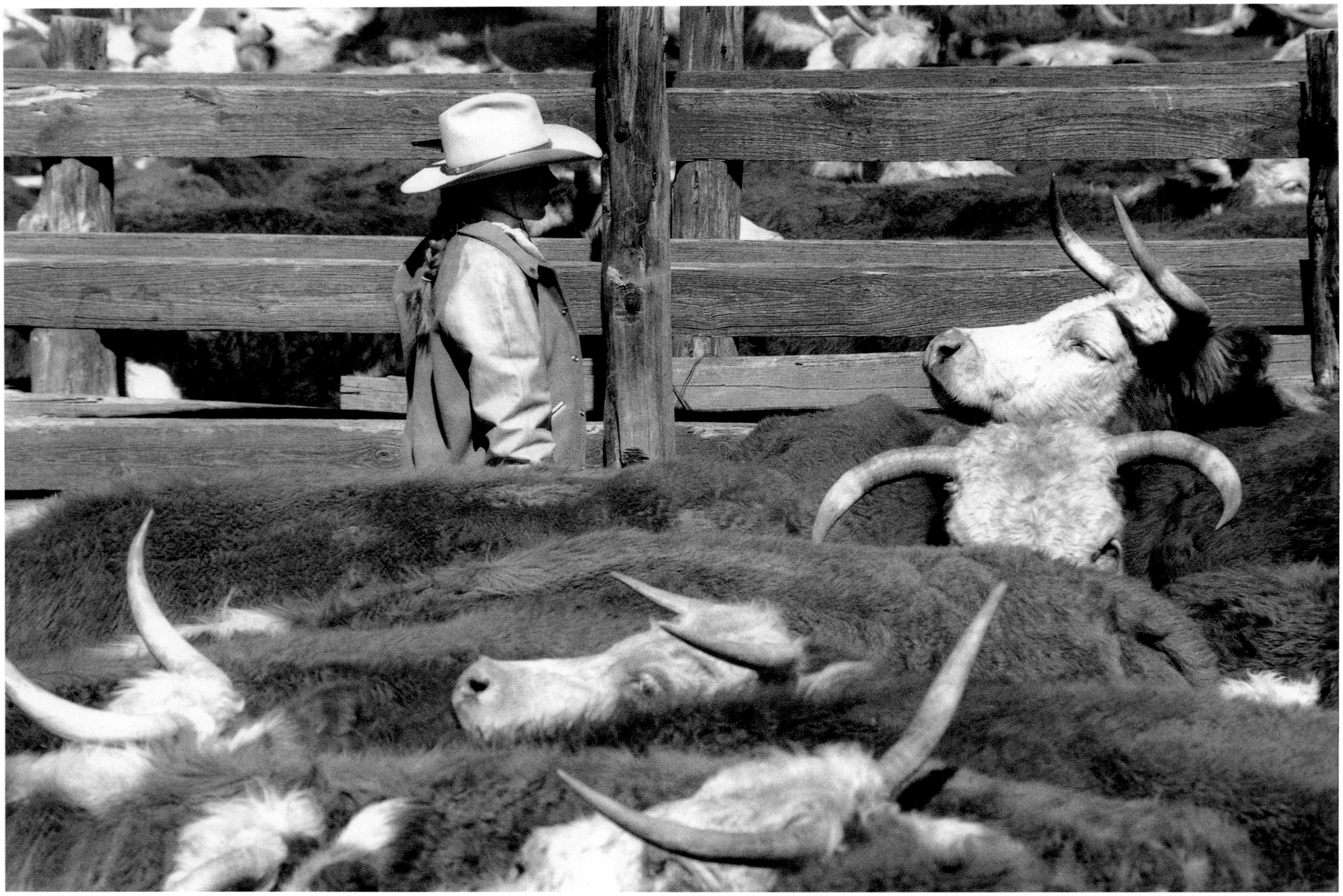

ANN HOLLAND DAUGHERTY, GAGE HOLLAND RANCH
1988

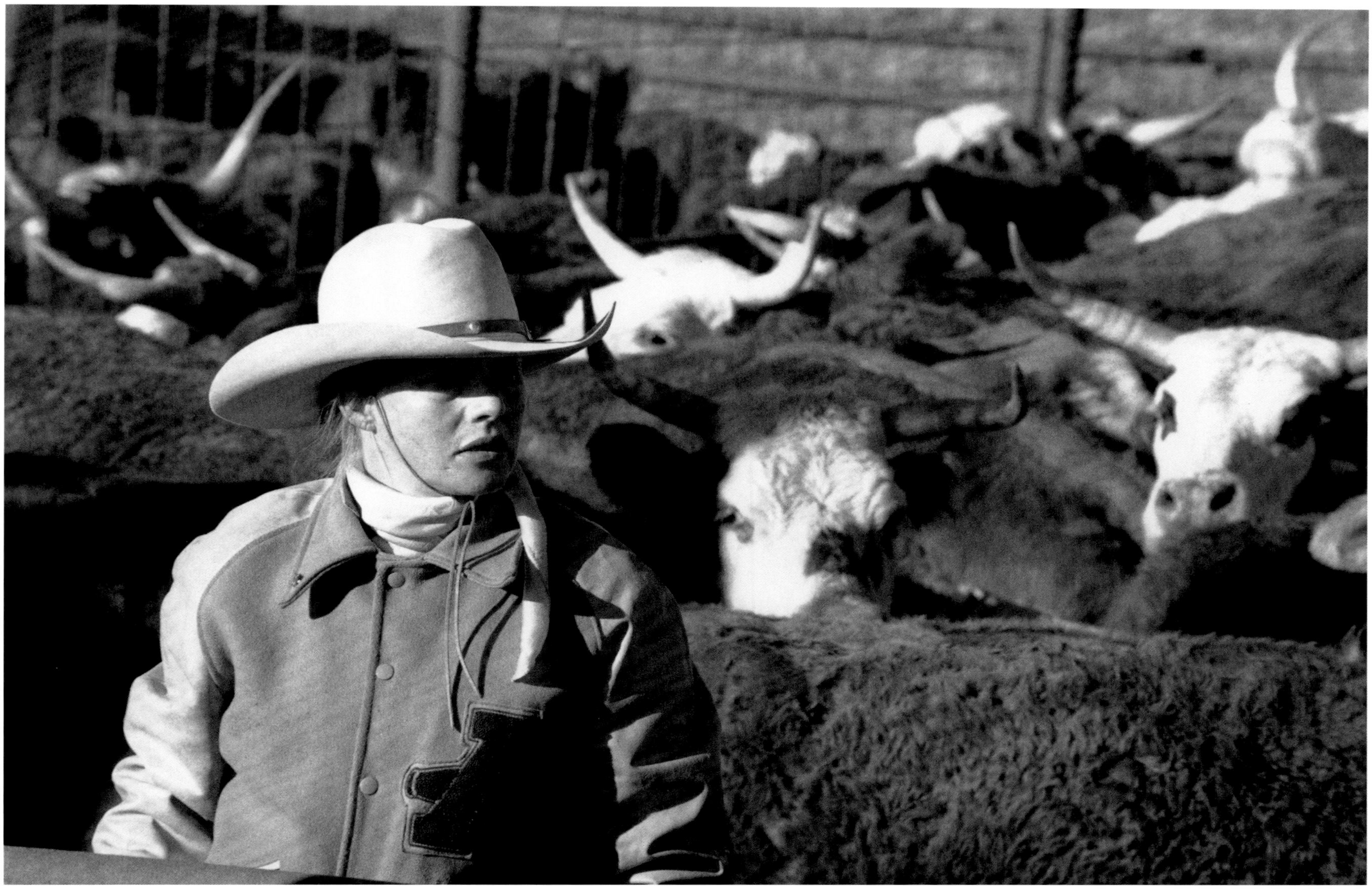

ANN HOLLAND DAUGHERTY, GAGE HOLLAND RANCH
1988

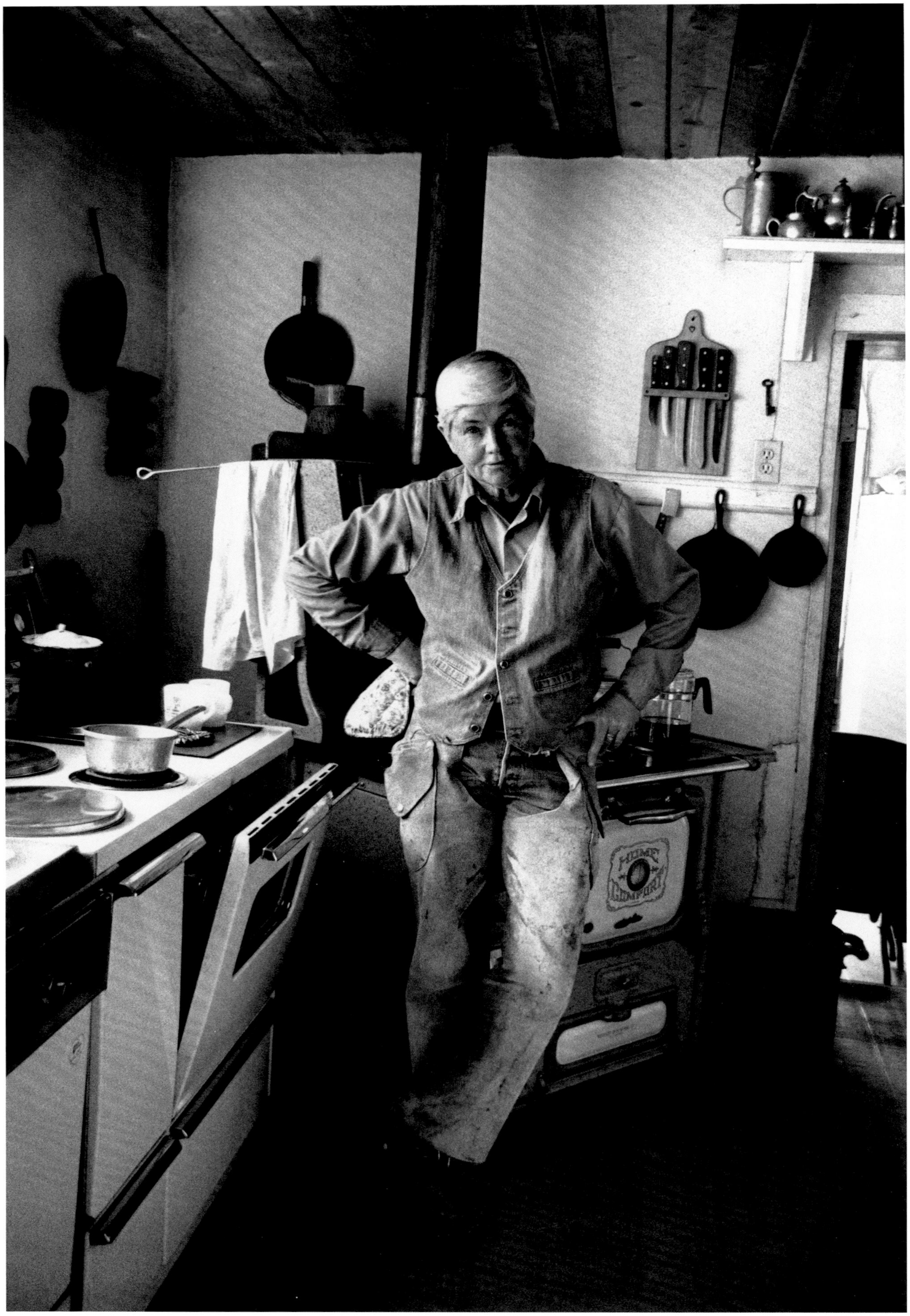

GRETCHEN SAMMIS, CHASE RANCH
1986

Gretchen Sammis and Ruby Gobble

Chase Ranch, Cimarron, New Mexico

Along the banks of Ponil Creek, in northeastern New Mexico, in a rugged piñon-studded canyon running out to meet the sprawling high plains, a former schoolteacher is ramrodding a ranch that has been in her family for four generations. Her great-grandfather, Manly Chase, founded this ranch 120 years ago when he traded a herd of wild horses for a thousand-acre piece of land. When he came west by oxcart over the Santa Fe Trail in 1867, he brought the mahogany bedroom suite that Gretchen still uses. "I was born in it, I sleep in it, and I expect I'll die in it," she says. In 1872, he built the old adobe house where Gretchen and her ranch foreman of thirty years, Ruby Gobble, live and heat each room with wood-burning stoves.

Gretchen Sammis, a tall, handsome, white-haired woman of sixty-eight, speaks quietly:

My grandparents raised me. They sent me off to the University of Colorado where I got an M.A. Then I came back here, ran my grandmother's dairy farm for five years, which was a disaster, then went to teach at the University of Wisconsin.

She gave that up after a year, moved back to Cimarron, and taught high school until 1973.

My grandfather had set the ranch up in trust before he died in 1964, but no one but me really wanted it. I had a good teaching job which I thoroughly enjoyed, and I loved every one of my kids. But, when the opportunity arose to take over the ranch, I took early retirement and here I am. I believe that very early in life an individual sets priorities that he or she is hardly aware of. Deep down, I always knew I would someday be running the ranch and taking care of it.

She, Ruby, and Bill Doerr, their hired hand, run 250 head of mother cows on 11,000 acres. "Bill's a good hand," says Gretchen. "He doesn't have any problem taking orders from a woman. A lotta guys do. Bill's different—he came from New Jersey."

"Sometimes when it is 20 degrees below, snow on the ground and calves coming, I know I had rocks in my head to give up teaching," Gretchen admits. "Or when the hay is ready to bale and it rains two inches, I wish I was somewhere else. But not for long. It is a challenging and rewarding life. One just rolls with the punches and hopes to go on forever." Times can be tough. Droughty or rainy summers or unusually hard winters, it doesn't matter, each is just as bad in its own way. So to help survive, Gretchen and Ruby, like many other ranchers, have diversified. They sell hunting leases on the ranch and also sell piñon firewood.

Ruby Gobble, a short reflective woman of sixty-four, grew up with a rope in her hands on a ranch near Wickenburg, Arizona. "My Dad died at sixty and Mom lived for another twenty years. They raised five of us kids out there on that miserable desert during the depression. It was tough, but they made sure we were fed and clothed." Although close to both her parents, Ruby's preference for an outdoor life bonded her to her father. He taught her to break horses, rope and then trick rope; at nineteen she went into roping professionally. In 1950 she was a runner-up for Women's World Champion Calf Roper; in 1951 and '53 she won the Women's World Champion Team Tying and in 1952 she won the Women's World Championship in Ribbon Roping.

Ruby is a jack of all trades. She can shoe a horse, cut a calf, and artificially inseminate cows. "I maintain the ditches, roads, and dams with the backhoe and 'dozer and do all the welding to repair the equipment. And I build gates and panels when they're needed." She's a hand! And she can cook, too.

How did Ruby get involved in the Chase Ranch as foreman?

Well, I needed someplace for my horses. I had mares, and Gretchen had a good Thoroughbred stud. I came over to ask her if she'd like to go into a partnership on the horses. Well, she had my horses here for several years before I moved over here to run it for her while she was teaching. And now she can't get rid of me—I'm too old to start over.

An old family tradition of true western hospitality continues, for at Thanksgiving and Christmas the old adobe house comes alive with music, singing, and a full table of friends and acquaintances.

Asked what they think outsiders and city folk ought to know about ranching that the environmentalists don't tell them, Ruby responds, "They need to know that it's not all glamour. They seem to think it is. They figure that it isn't any work. You just turn the bull out with the cows and that's that. It is a long way from being that simple." Gretchen adds,

They need to know that there is a hell of a lot of responsibility, trying to figure out what to do. Do we move the cows now or later and so forth. They should know that the only reason people ranch is because they like it. You sure can't do it for money! My God, look what you've got invested in it. If you get any return at all, it is a very, very small percent of your investment. Nobody in their right mind would put all their capital in something for a one or two percent return. True, but there are a lot of other rewards.

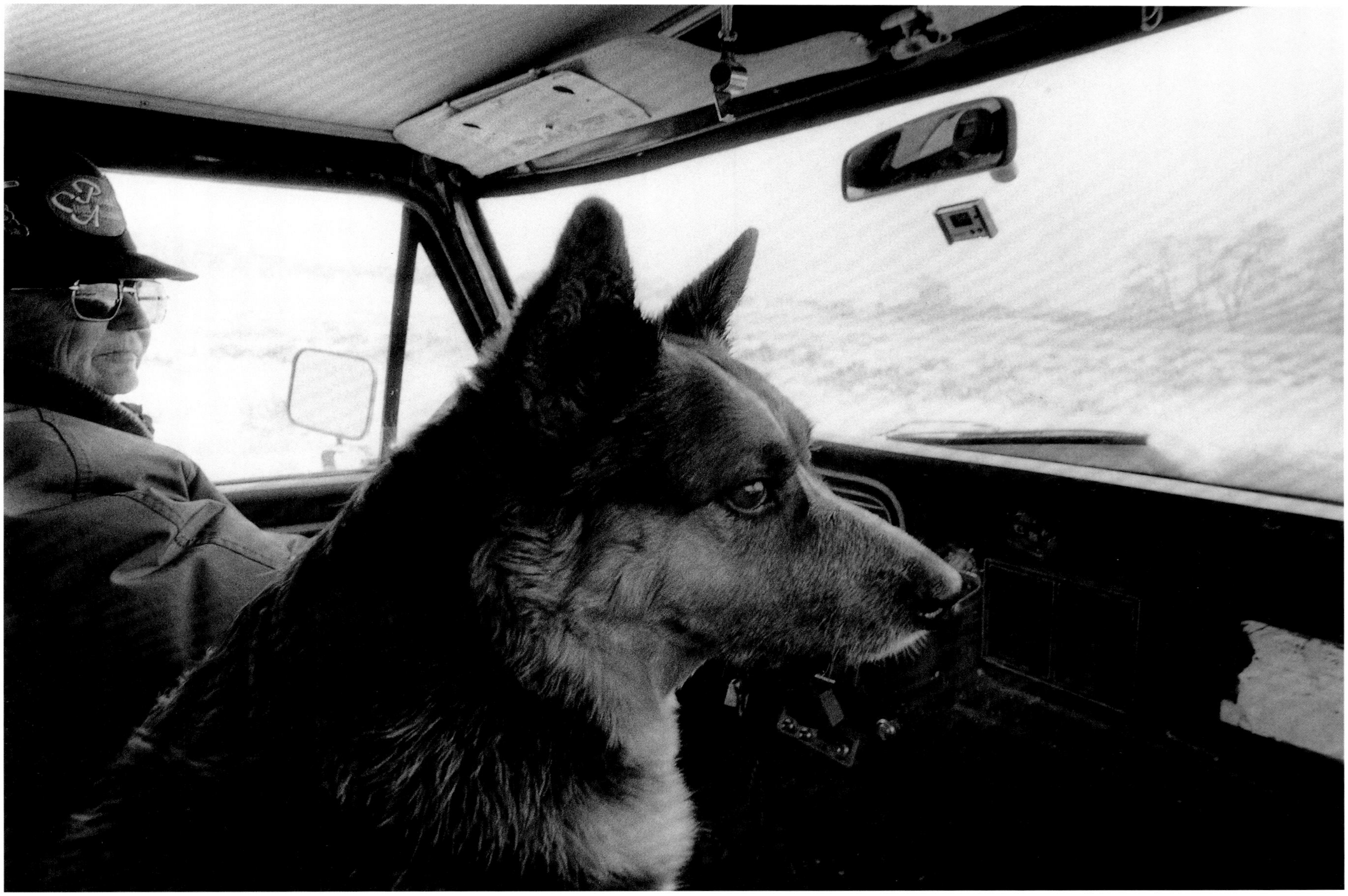

RUBY GOBBLE, CHASE RANCH
1986

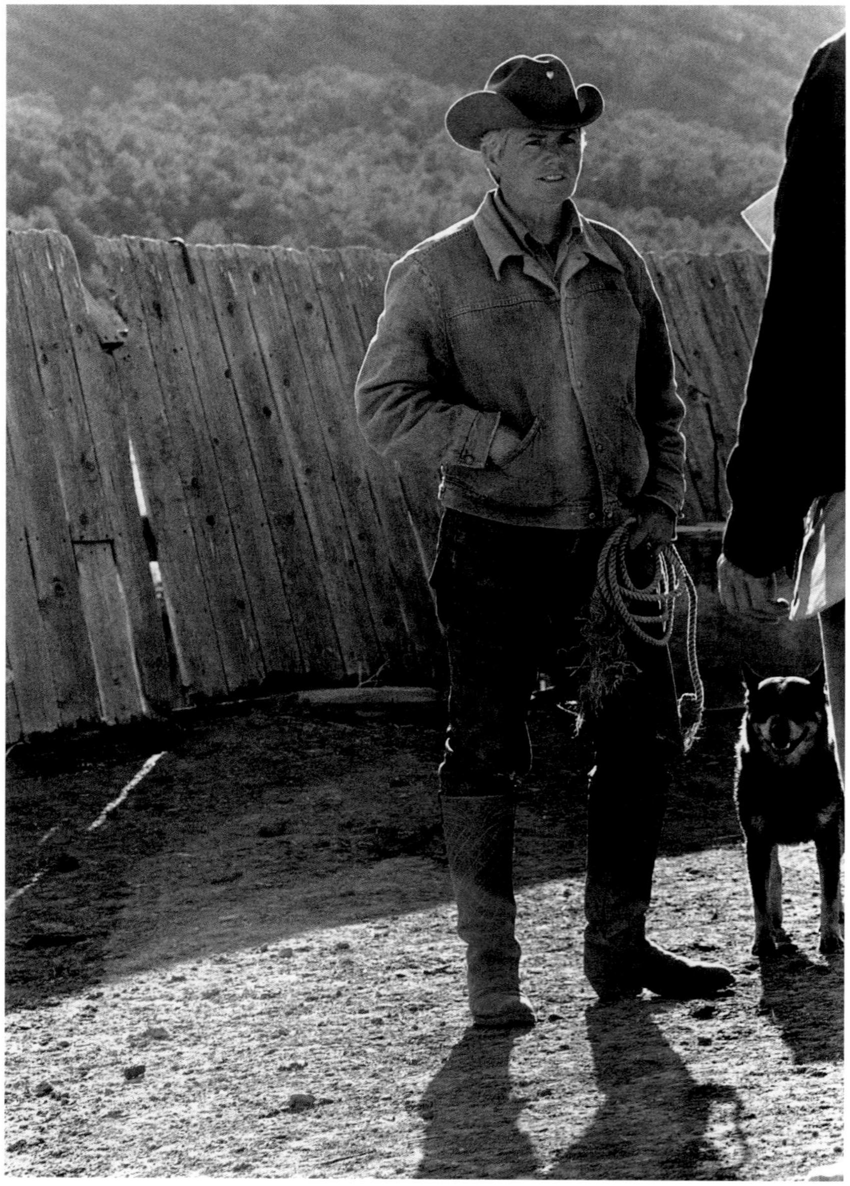

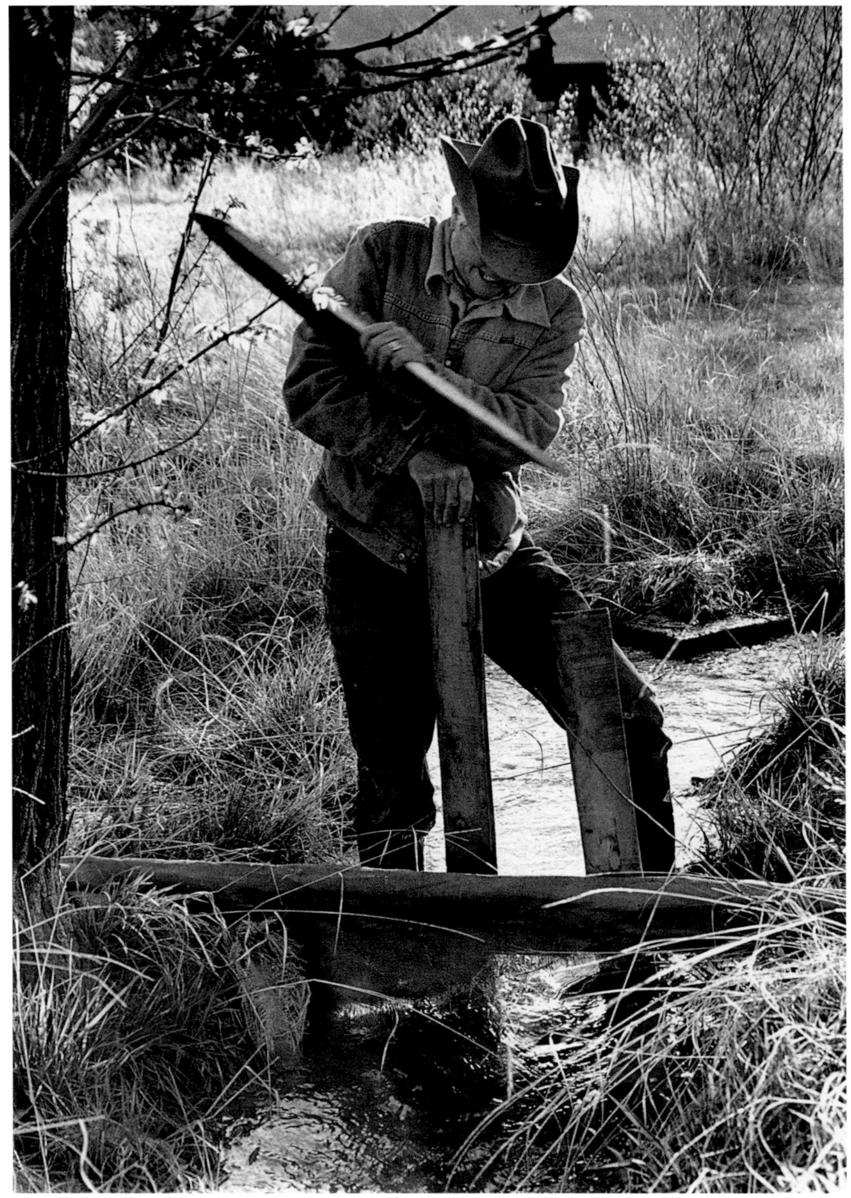

GRETCHEN SAMMIS, CHASE RANCH
1986

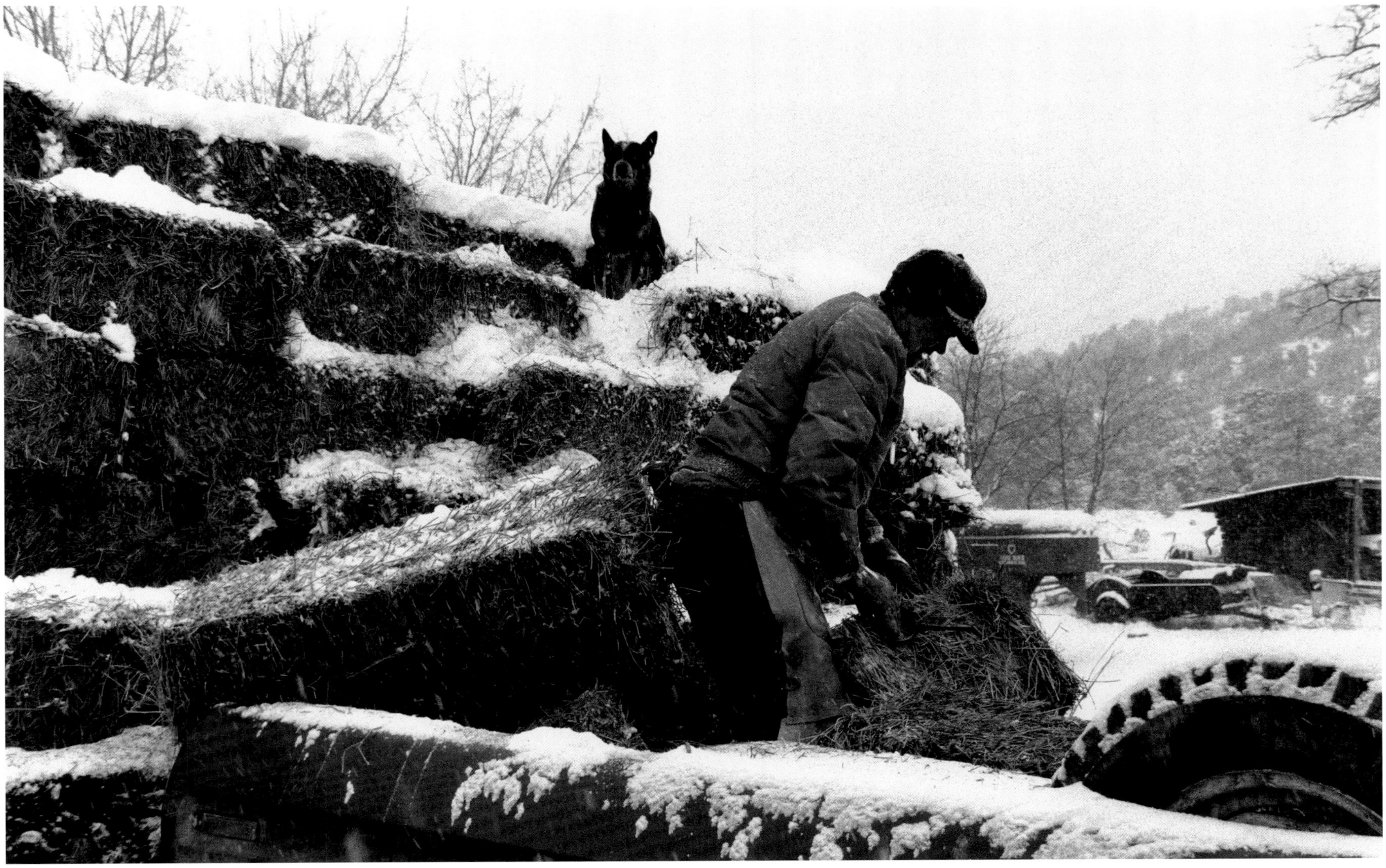

GRETCHEN SAMMIS, LOADING HAY, CHASE RANCH
1987

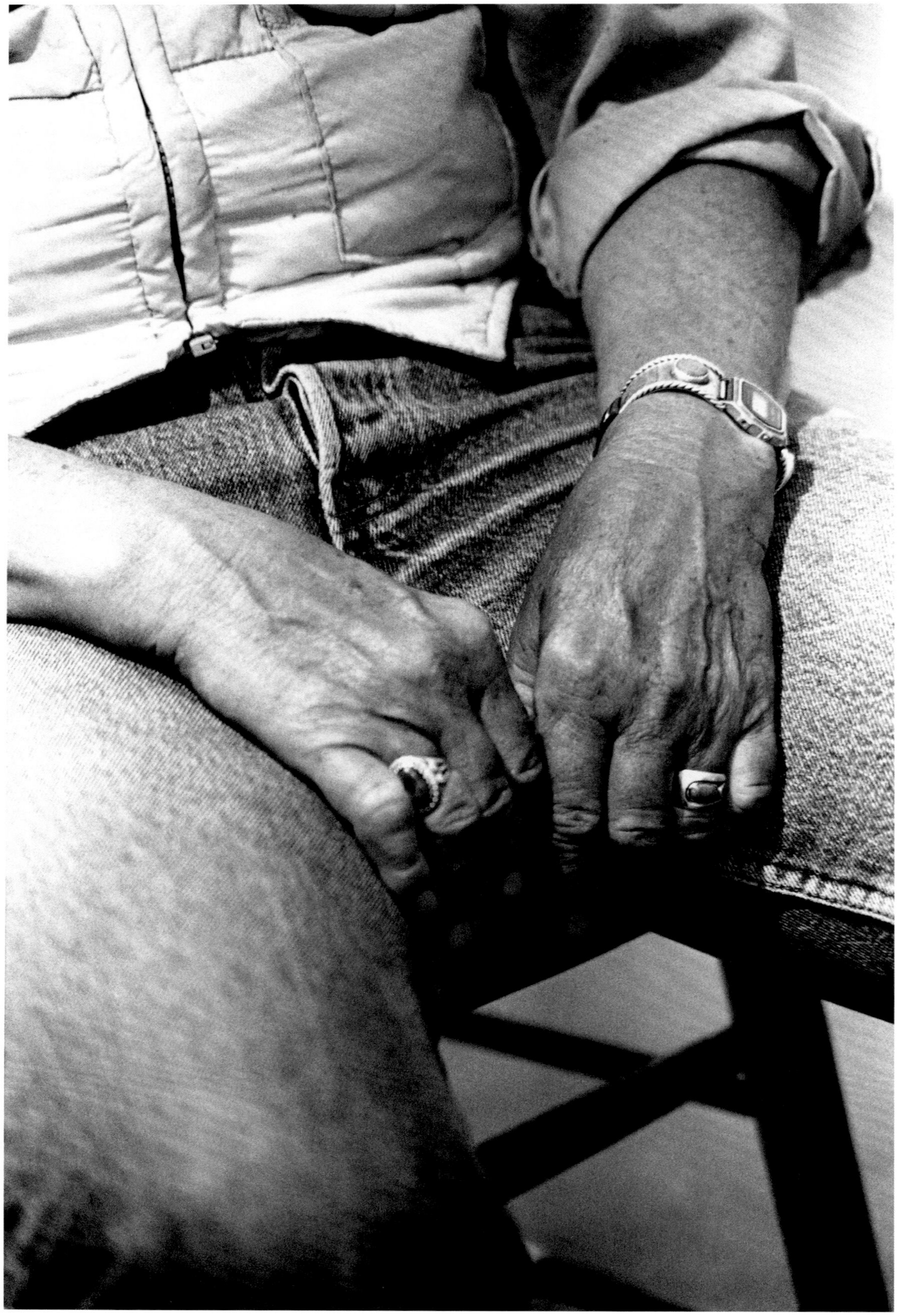

RUBY GOBBLE, CHASE RANCH
1986

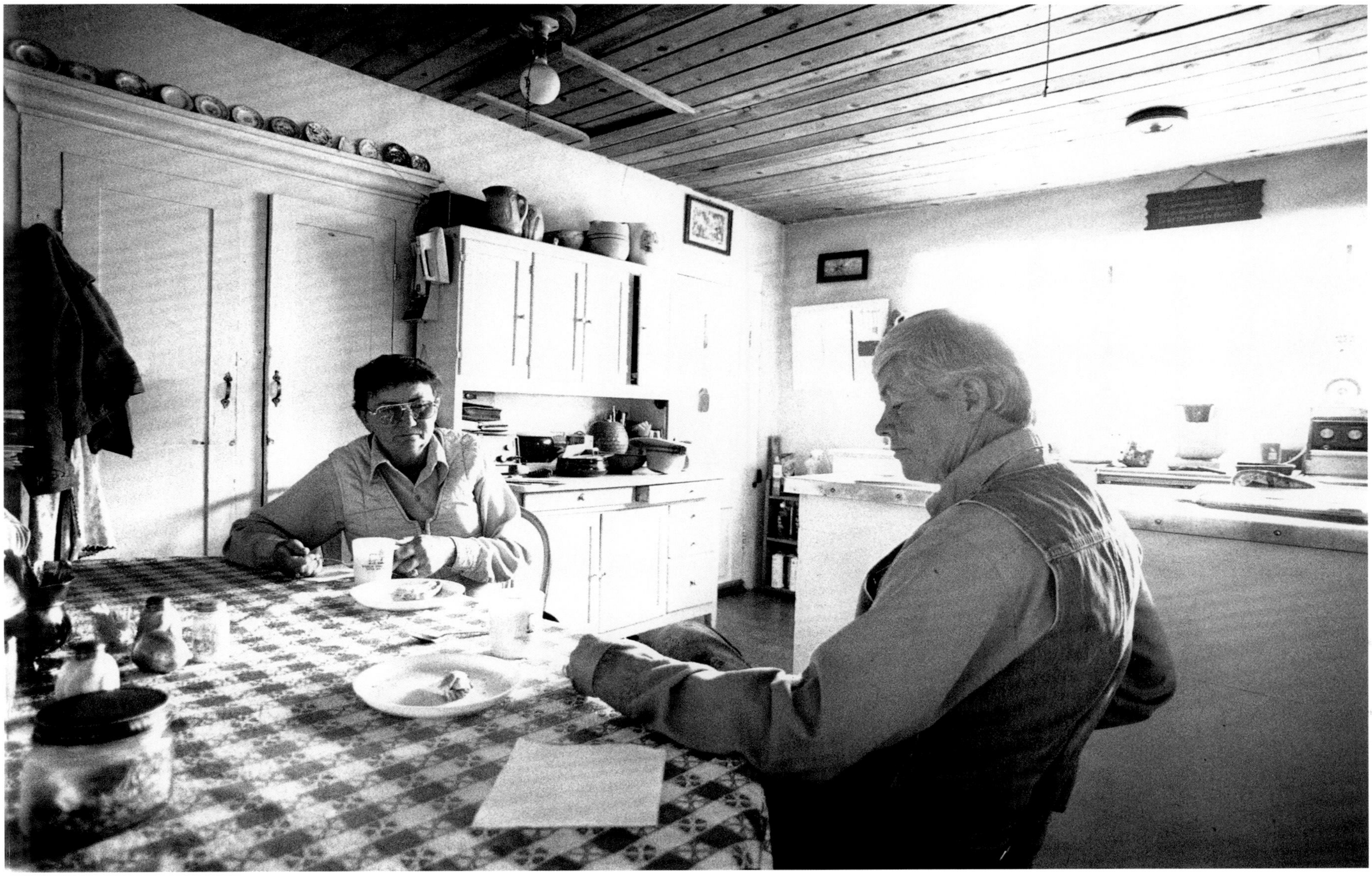

GRETCHEN SAMMIS AND RUBY GOBBLE, CHASE RANCH
1986

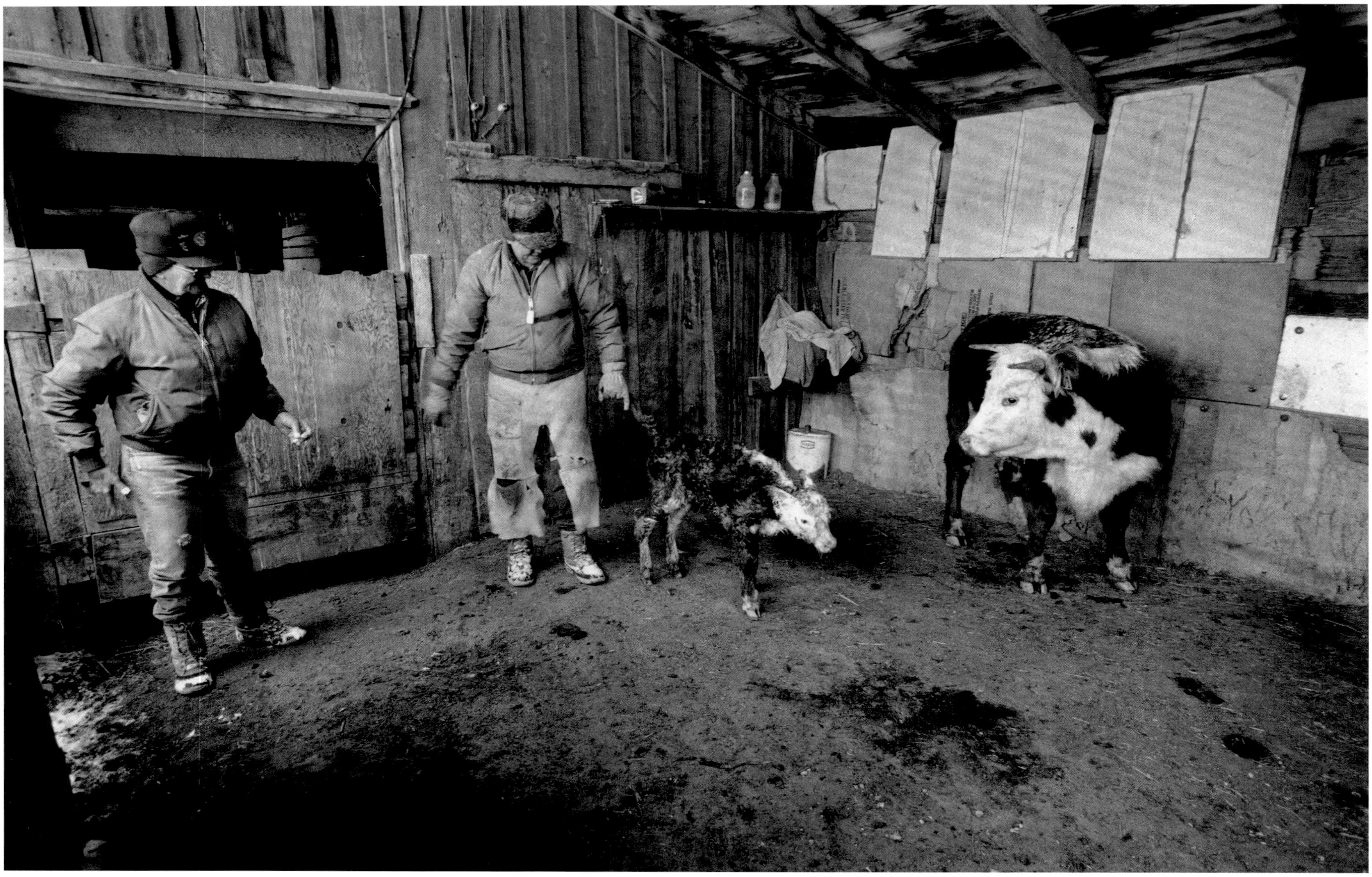

GRETCHEN SAMMIS AND RUBY GOBBLE, CHASE RANCH
1986

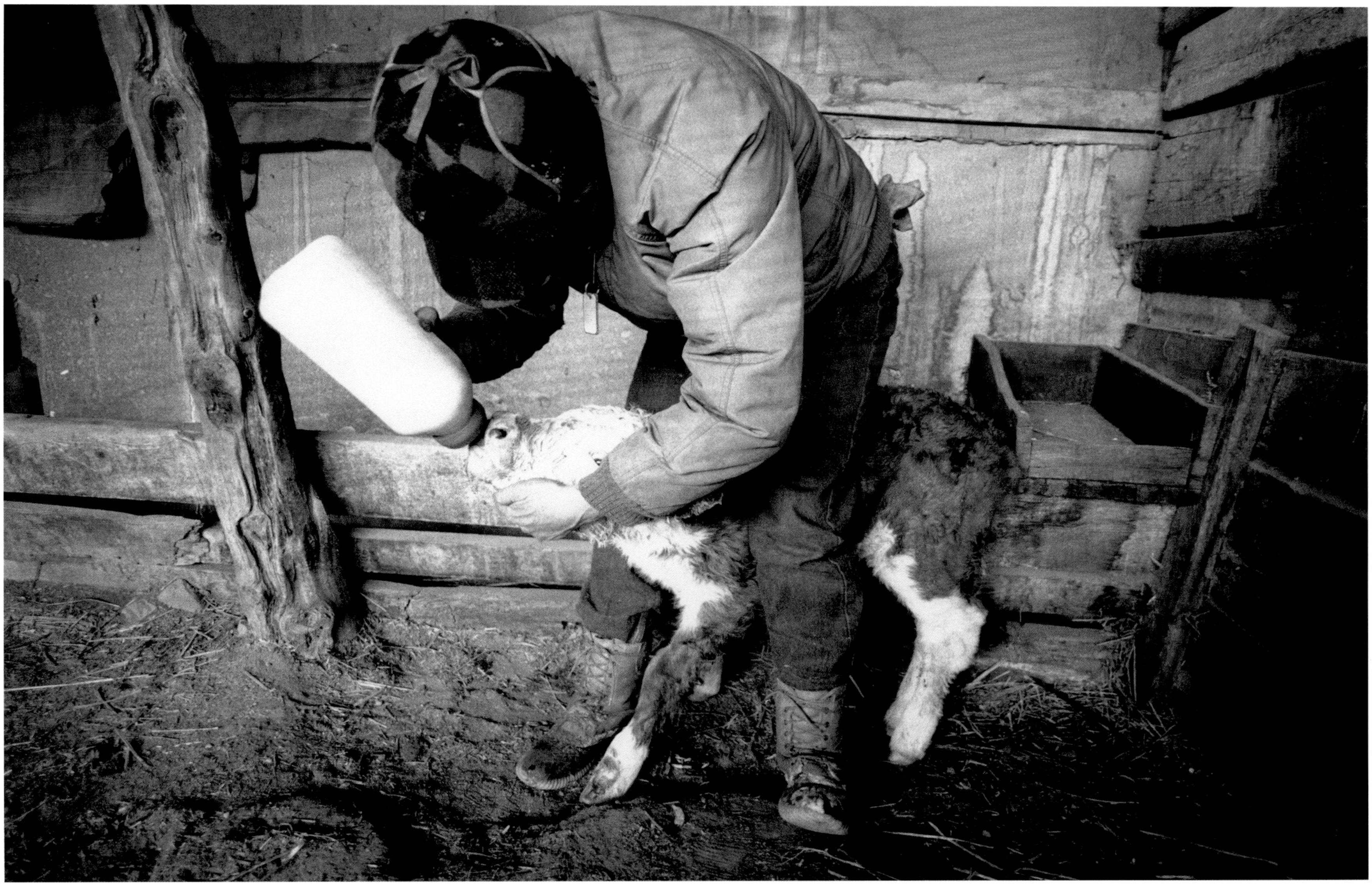

GRETCHEN SAMMIS, BOTTLE FEEDING AN ORPHANED CALF, CHASE RANCH
1986

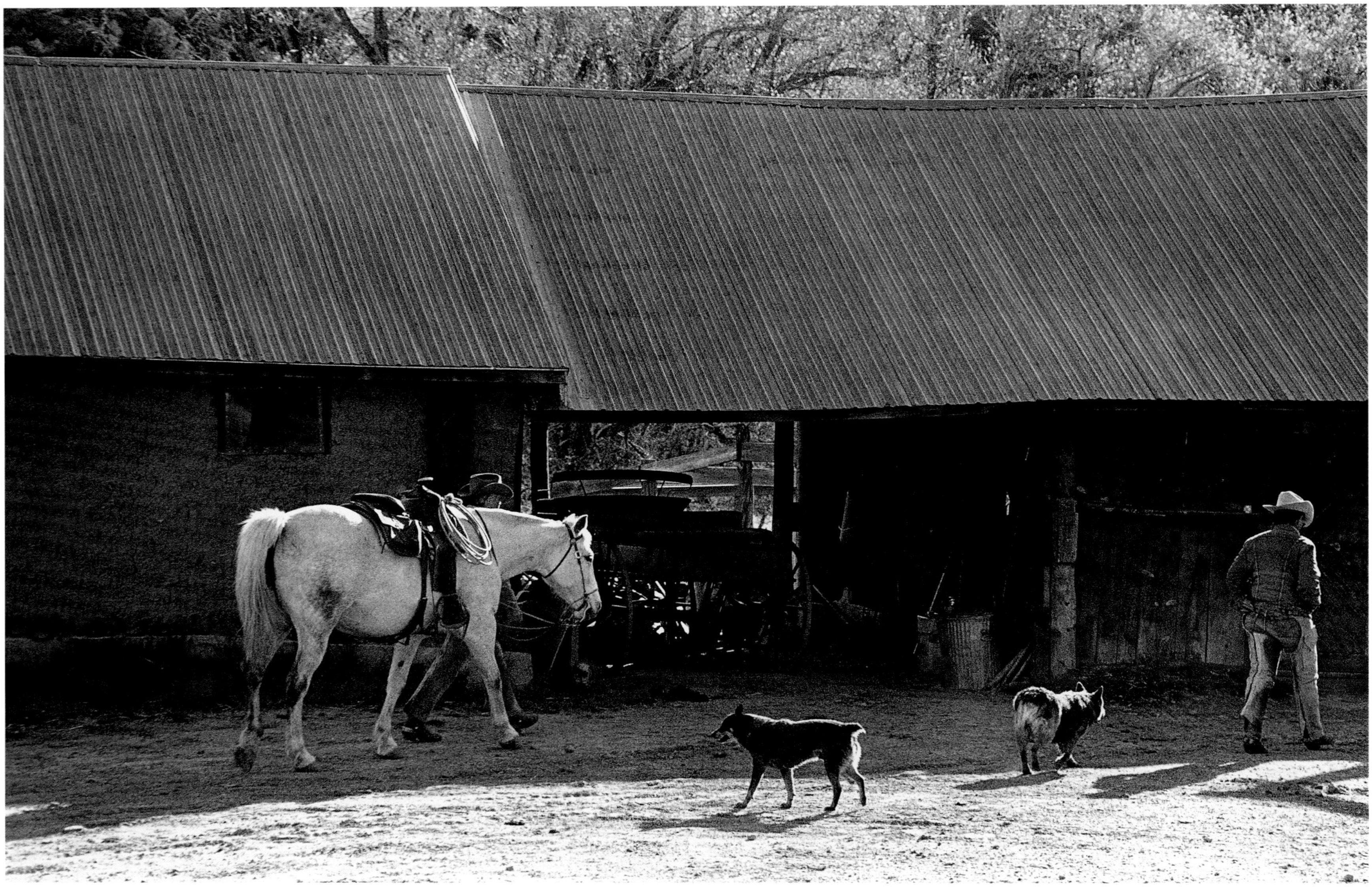

GRETCHEN SAMMIS AND RUBY GOBBLE, CHASE RANCH
1986

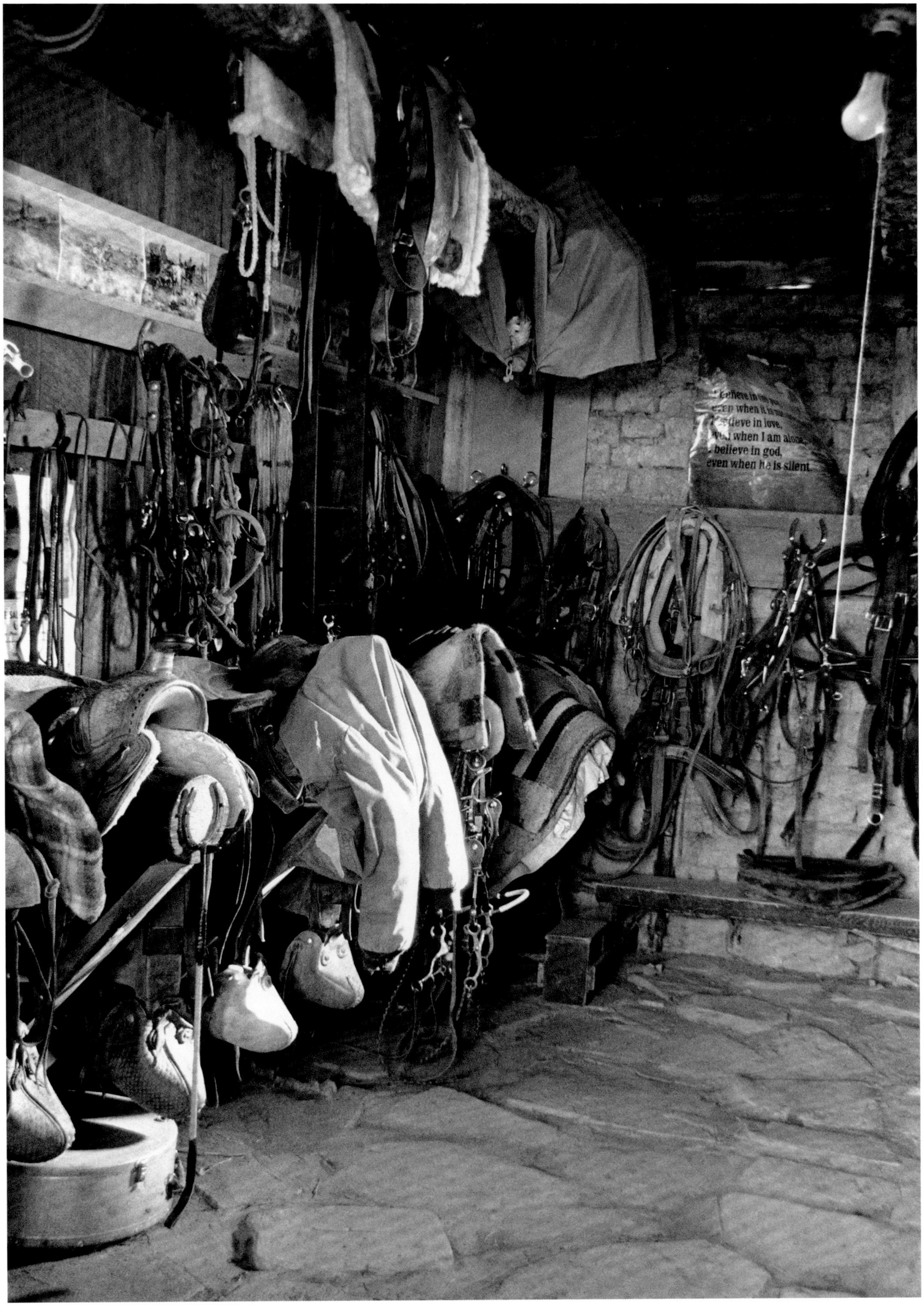

TACK ROOM, CHASE RANCH
1986

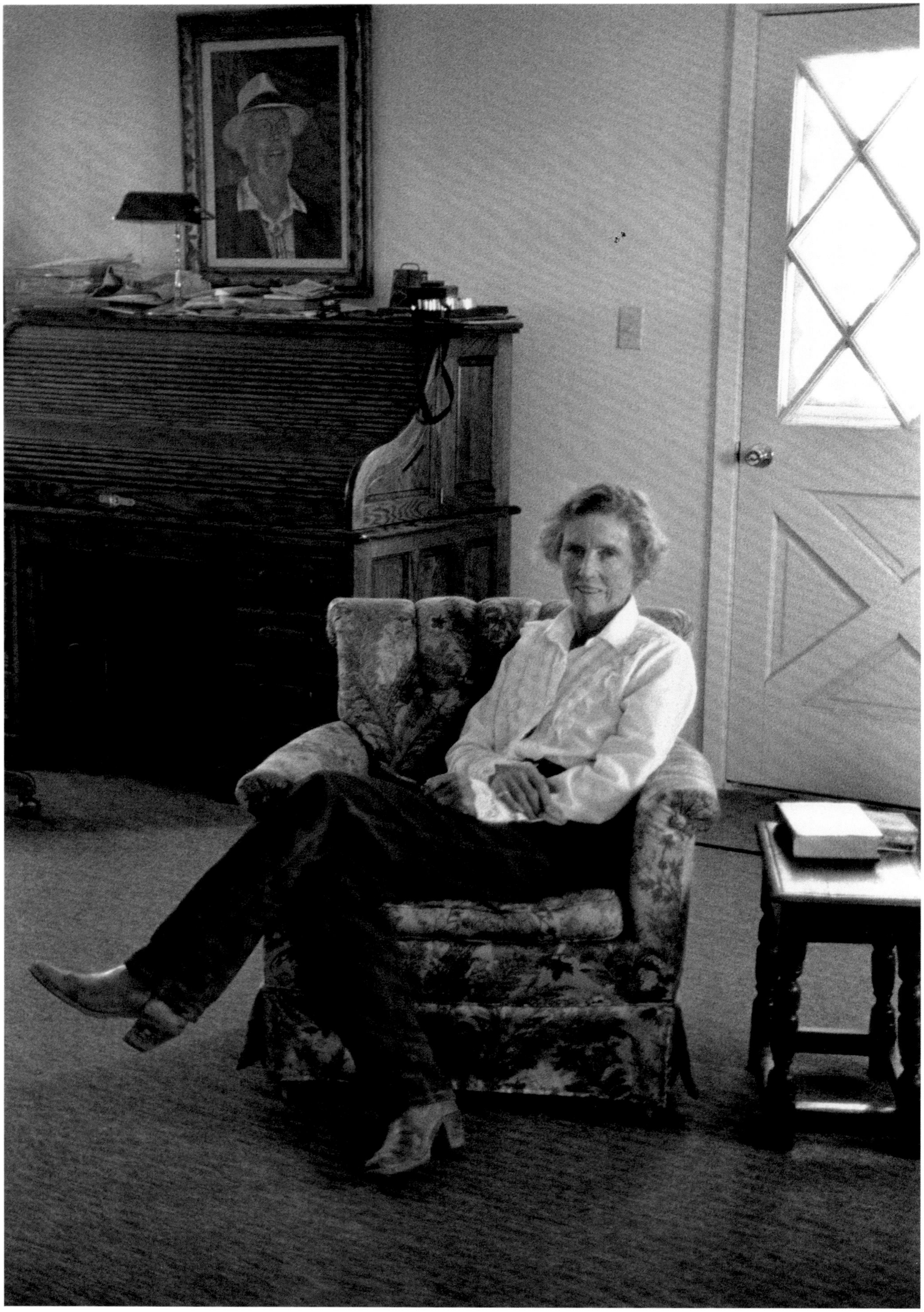

MOLLY FLAGG KNUDTSEN, GRASS VALLEY RANCH
1986

Molly Flagg Knudtsen

Grass Valley Ranch, Austin, Nevada

Getting there across the vast, barren expanses of Nevada was an education in itself. From I 40 at Battle Mountain we headed south across a near desert of dirt. In the distance a string of horses, probably wild horses, stirred up lines of dust that hung above them to mark their progress. It was a long drive. Finally, we entered a good, grassy valley, arriving at an orderly ranch of numerous buildings and corrals. We're welcomed by a warm and gracious lady whom I already knew to be a former debutante.

A debutante rancher? You bet! She's ninety-five pounds soaking wet, a dynamic, witty lady in her late sixties. She had just finished vaccinating two hundred heifers that morning. She helps with the calving but does all the branding, vaccinating, the earmarking and castrating herself.

The hired men push the calves into the chute and table where Molly does the work — "I wouldn't turn my cows over to anybody else for anything under the sun." She handles eight to nine hundred calves a year. Additionally, she raises about a thousand head of commercial Herefords and keeps a herd of one hundred purebred Herefords on the beautiful and well-watered high-desert 9,000-acre Grass Valley Ranch.

Molly was the only child of an heiress and an architect. He died when she was seven, and her mother married a U.S. Army colonel and they traveled extensively. "I was raised all over the globe." As a young socialite, she made her debut before the king and queen of England and was involved in the international horse circuit. Then, in her early twenties, she became one of three licensed women horse trainers in America, which brought her to Reno in the early 1940s to help train race horses on a friend's ranch.

During this time she met the attractive but unpredictable Richard Magee, who was considerably older. They married and lived on his Grass Valley Ranch. She, with typical drive and determination, threw herself into learning everything she could about Nevada: its history, peoples, educational system, and about the ranch itself. Her first animal experiences were with horses, but as she says, "you find no shortage of people who like horses, but there seems to be an aversion to loving something that will be eaten. I happen to like cows. I lived for twenty-six years on a horse ranch and I starved to death for twenty-six years. I'm not raising horses!"

During those first twenty-six years on the ranch they never hired a buckaroo. Molly did all the doctoring, branding, turning out, gathering, and working of the Grass Valley cattle herself. Then Dick went to California and never returned. "Dick was a bachelor when I met him and he was a bachelor after twenty-six years of marriage. When he left me in solitary splendor, I was delighted, but it was difficult."

Her interests have always been diverse: "I grew concerned about the churches in the area, many of which were crumbling and abandoned," so she became involved in their restoration. Because she also was worried about the rural youngsters and the kind of education they were getting, she considered running for the school board, but others urged her to run for the university (of Nevada) system's board of regents. In those days (1960), it was a male-dominated board to which no woman had ever been elected. Molly recalls, "At that time, there were eight hundred women students. I've never been a feminist, but that challenge was like a spur in my side. I've done things all my life that women weren't supposed to do!"

She campaigned on the platform "You need a woman on the board of regents," won by a landslide, and remained on the board for eighteen years until she retired in 1980. She supported, from the first, the effort to establish the University of Nevada Press and helped to establish the department of anthropology and supported the Desert Research Institute at the university. In addition, she has written two books about Grass Valley: *Here Is Our Valley* and *Under the Mountain*, as well as a number of monographs.

About a year after Dick Magee moved to California, she married Bill Knudtsen, a horse trader in the style of the old-time mustang runners. "It's been a happy marriage," Molly says. She continues, relating a story about a hired man who made a derogatory remark about Bill which she didn't like. Right then and there she fired him, " 'Go roll it up.' He was the most surprised man you've ever seen," she chuckles. She's not afraid: "I've pulled a gun on people if at night they come in the yard. I've no compunction about defending myself or the ranch."

Does she worry about anything?

Not much, except the takeover by the U.S. military of more and more of Nevada. They're condemning ranches — they're buying out this whole country here for testing, bombing, and everything. I am concerned that the ranchers' days are numbered. The wilderness people and the nature lovers will probably get us off the ranches. I have a good relationship with the BLM. They don't bother me. I'm just as interested in preserving this ranch as they are, probably even more so. I take a tremendous interest in the natural forage on the ranch, and they are aware of that and like working with me.

Does she feel that she is disadvantaged by her age and slightness? "Not one bit! It used to terrify me to think of getting old and feeble, but you know, . . . I seem to have gotten old but I haven't grown feeble yet."

An incident occurred when I was there. We'd gone out with her to see her cows while she worked them. In the course of this, Timmy, her horse, jammed a foot in a gopher hole, catapulted Molly over his shoulder and then fell on her. She had a couple of broken ribs and a lacerated liver. It took the ambulance nearly three hours to pick her up, get her to town, and put her on a plane for the Reno hospital. It took her quite a while to heal up from that one. Perhaps she never did heal completely. Not long after she sold off her cattle, and in spring 1995 Grass Valley Ranch was sold to her neighbors and longtime friends.

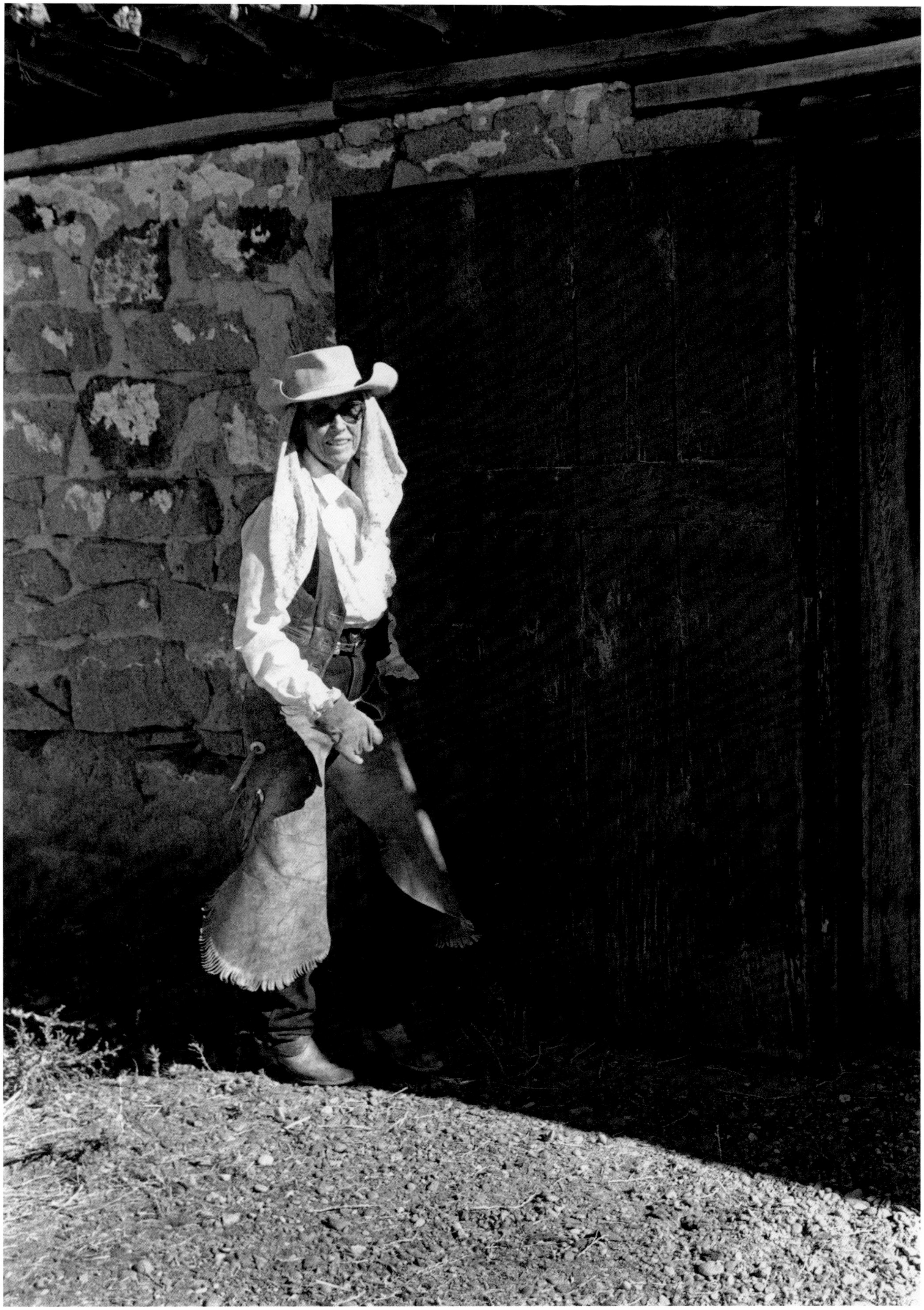

MOLLY FLAGG KNUDTSEN, GRASS VALLEY RANCH
1986

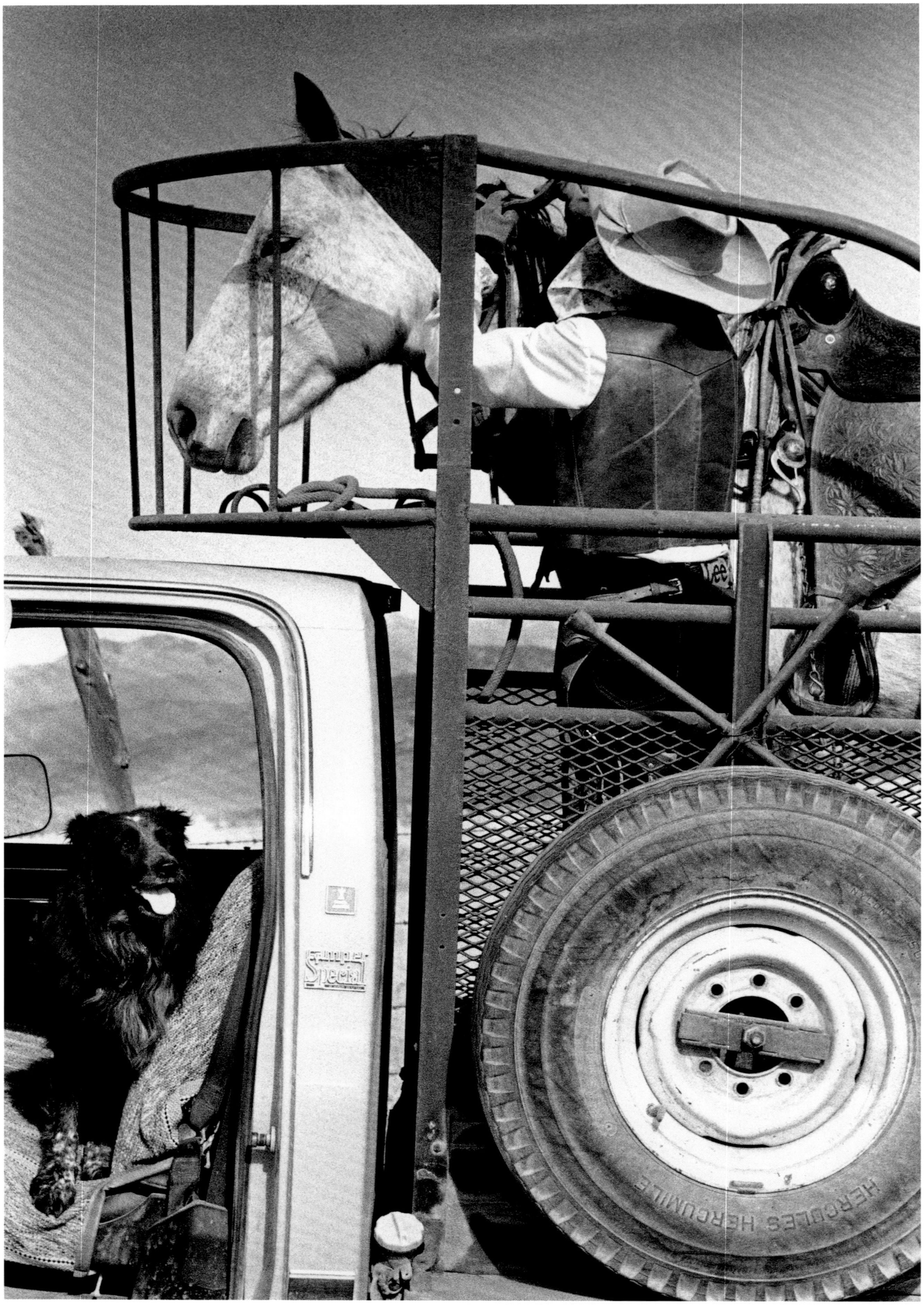

MOLLY FLAGG KNUDTSEN, GRASS VALLEY RANCH
1986

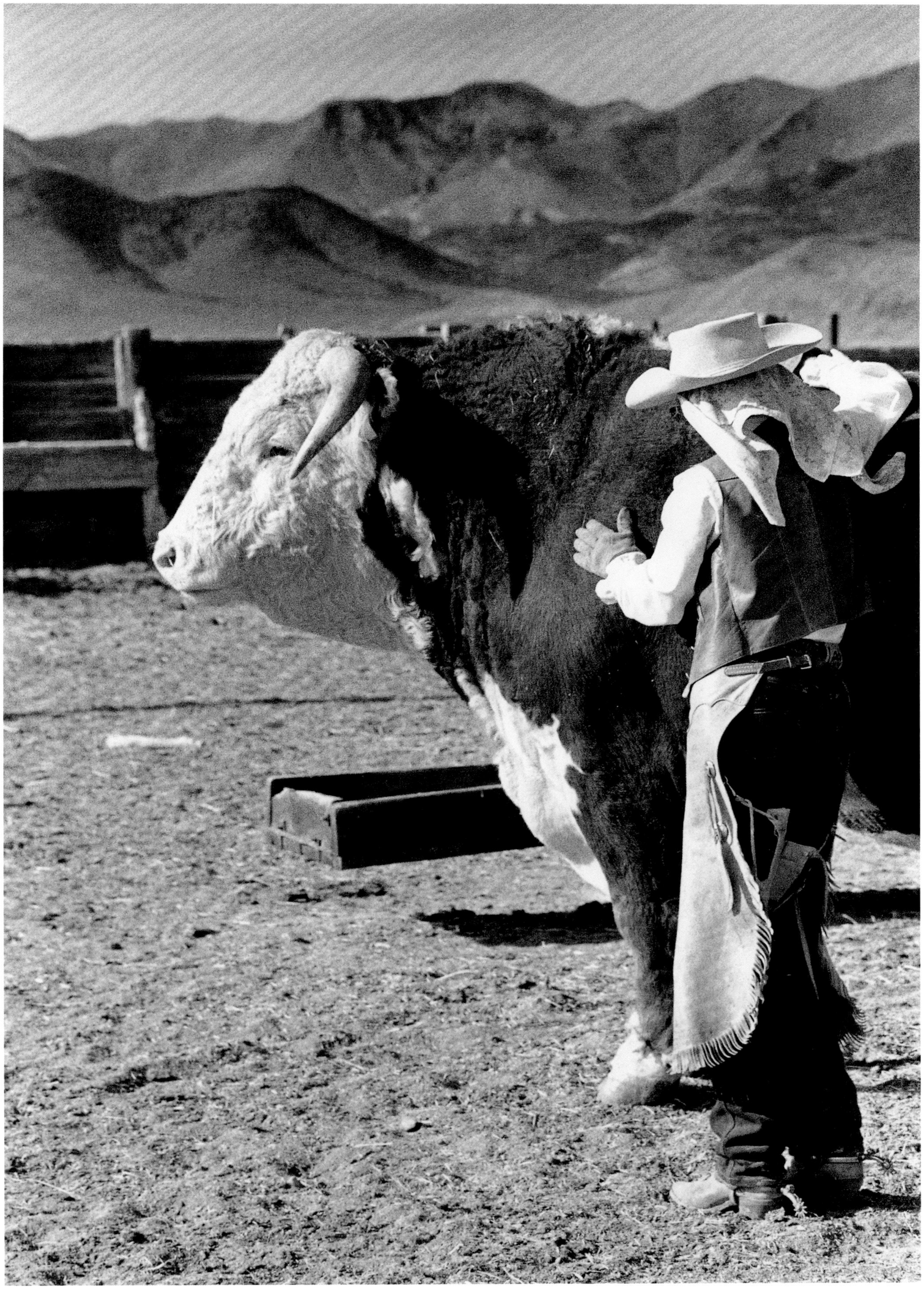

MOLLY FLAGG KNUDTSEN, GRASS VALLEY RANCH
1986

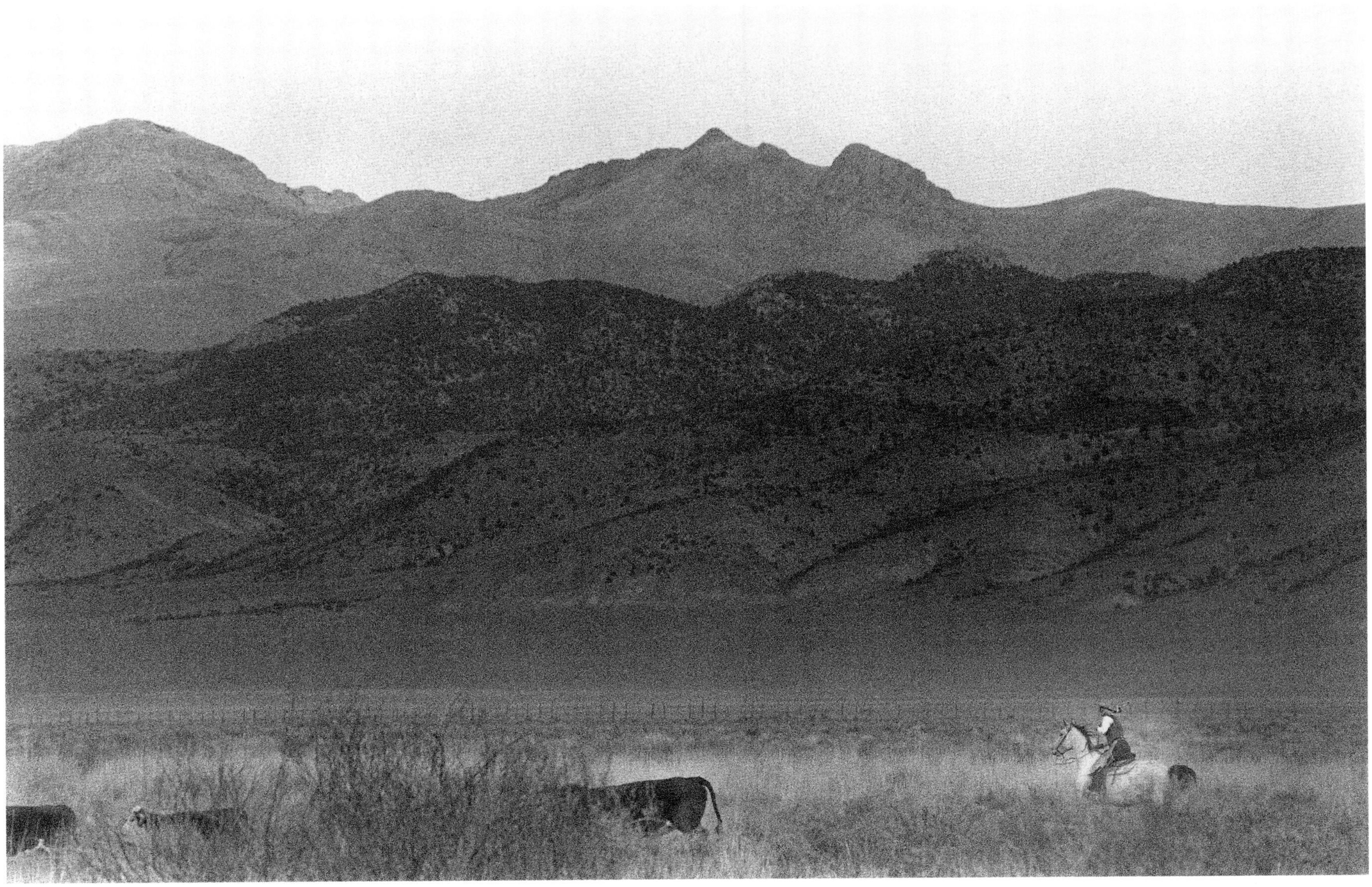

MOLLY FLAGG KNUDTSEN, GRASS VALLEY RANCH
1986

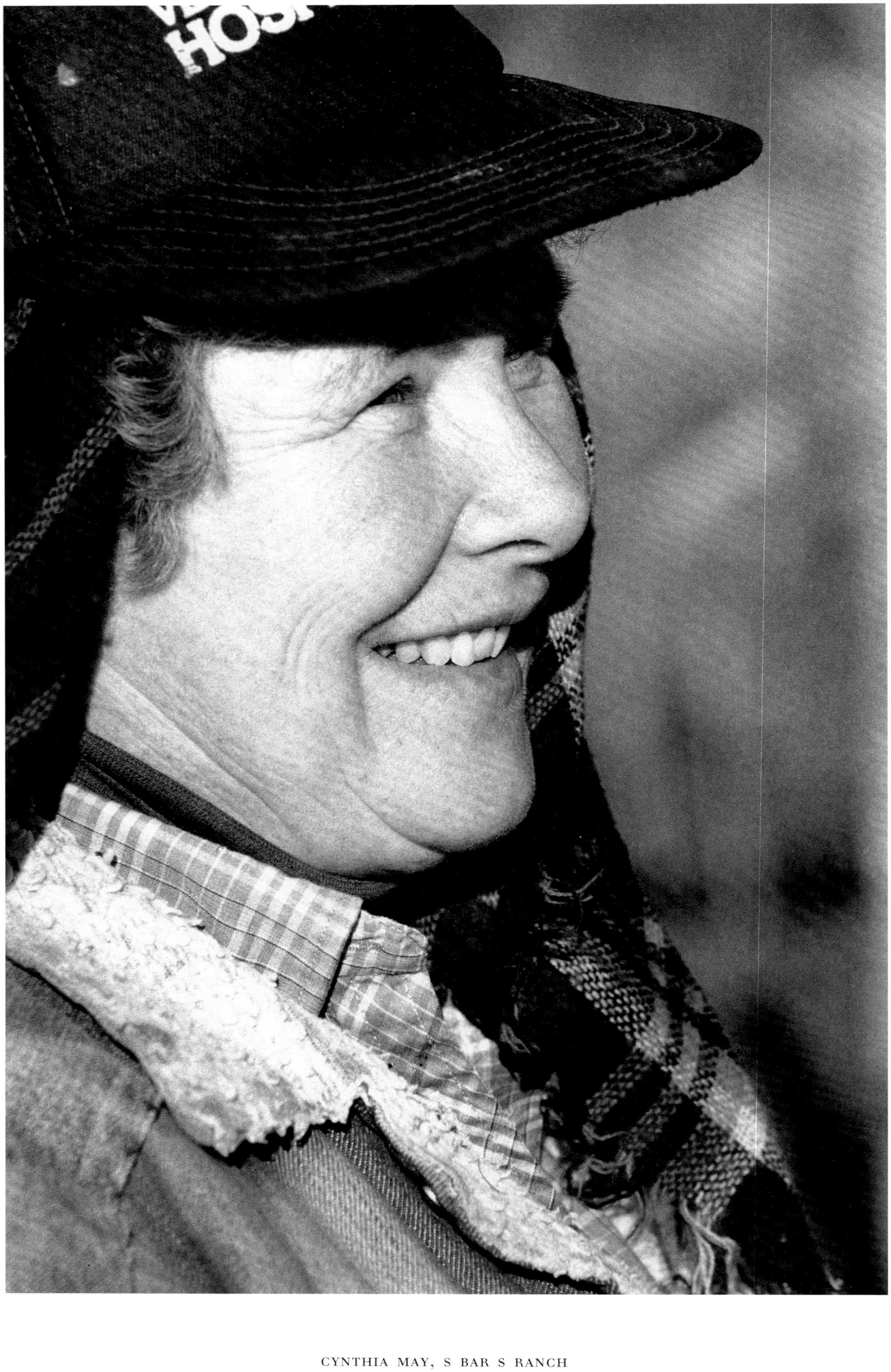

CYNTHIA MAY, S BAR S RANCH
1987

Cynthia May

S Bar S Ranch, Steamboat Springs, Colorado

A wholesome woman of sixty-one with a wonderful smile, Cynthia walks out to greet me on a frosty late afternoon in March. The ranch is situated about ten miles north of town in the lovely, fertile Elk River Valley. Her husband's family has been here since 1899. Husband Bill is a cowboy poet who spends a good deal of his time in front of an old typewriter composing epic-length poems. He hopes eventually to be able to support his family on income from writing and reading his verse.

Cynthia grew up on her family's ranch in northeastern New Mexico in "great, great cattle country—when it rains!" There was no phone on the family ranch nor were there any hospitals in the area. She was born in a private home where the doctor came to deliver babies. As a girl she went to a small country grade school with one teacher and twenty-one students but only two children in her class. Then, like most ranch teenagers of that era, she attended high school in the nearest town by boarding at someone's home during the week and returning to the ranch on weekends.

My dad always had us helping with the cattle work but he wouldn't let us drive tractors or anything like that. I remember the first time [on the ranch in Colorado] I drove a team of horses pulling a hay rake. They were a gentle, old team, but the men told me to watch out for the sloughs out in the field. So every time I went around a turn with the team, my heart was going "Boom, boom, boom!" I was scared to death because I didn't know the terrain.

Despite asthmatic bronchitis and a bad knee, Cynthia spends most of her time in her beloved outdoors. She invited me to come see her mount, "Old Grey Henry," which turned out to be her grey Fordson tractor. Because of her health problems, she rides this around the ranch, often with a hired hand astraddle the hood of it so he can rope, if necessary, or hop off easily to open gates or irrigate. Even though she loves it, the ranch is a lot of work for her.

Cynthia was sixteen when she and Bill were engaged to be married.

When I was nineteen I just knew it all. Now I don't know nearly as much! Bill and I got married when he got back from Germany, and we went to live and work on his dad's ranch because he threatened to sell it if we didn't. Dads have a way of doing things like that. When we were married I had this dream, this vision of all the kids living on the ranch helping us. That didn't work out. We just didn't know that times were going to get so hard, and I never realized that four kids is too many to have on a ranch the size of ours.

Cynthia and Bill May have a small outfit of 1,600 acres and government grazing permits that cover about 30,000 acres, much of it steeply sloped, timbered, and rocky. Cynthia has some strong opinions about ranching and the government:

I don't think the government should be buying any more private property, and I sure wish they would stay out of the market. If people would just leave things alone then the market would likely either go up and stay or go down and stay, but these wild fluctuations are hard to take.

I don't know hardly anybody [in ranching] riding in Cadillacs, and they're having hard times making ends meet. This guy that we do business with, I was bellyaching about the cattle industry, I guess, and he looked at me and said, "Tell you what, would you trade that ranch out there for a high-rise in Denver?" I said, "No way!"

If an urban lady suddenly became wealthy, she'd be taking trips, buying a fancy car, clothes and I don't know what all. Do you want to know what I'd do if I had a bunch of money all of a sudden? First thing, I'd start with a good set of corrals and a nice pickup and a horse trailer and a good stock truck. I'd settle with that for starters.

Their three sons are married and hold down other full-time jobs and help when they can on the ranch. Their only daughter, Grace Ann, "was our top hand until she got married. She's on a picturesque ranch in Utah now."

The saddest thing in my life was the death of our horse Moon. Family deaths are terribly sad but usually those people have lived long and full lives. One time, Grace Ann was eleven or twelve. She was riding this wonderful horse. You could catch him in any pasture and then ride him off with just a string around his neck. Well, somehow, while Grace Ann was riding him, he ran into brush or something and punctured a big artery in his groin with a stick. He just literally bled to death right in front of her eyes. He was looking at us, wanting us to somehow help him. We couldn't reach one vet—they were all gone, and we just couldn't do a thing to help him. He was just a young horse, four or five, but was just one of those great horses. It was just awful. When you have a cow that butts you or a horse that kicks and bites you, well, you don't mind saying "Adios," but this horse, it was a tragedy.

Ranch life can be both hard and heartbreaking, but Cynthia May wouldn't trade it for anything:

It is a good life. What's really great is to go out and check the cows early, early in the morning. You can hardly even see them, but it's just the break of dawn and here comes the wild geese a-honking up Spring Creek Canyon and over on the other hill are coyotes howling while some over here are yipping back. There's something about the quiet and the natural sounds. Then in the evening, I like coming in from irrigating and listening to the "whoooooo" of the Wilson snipe.

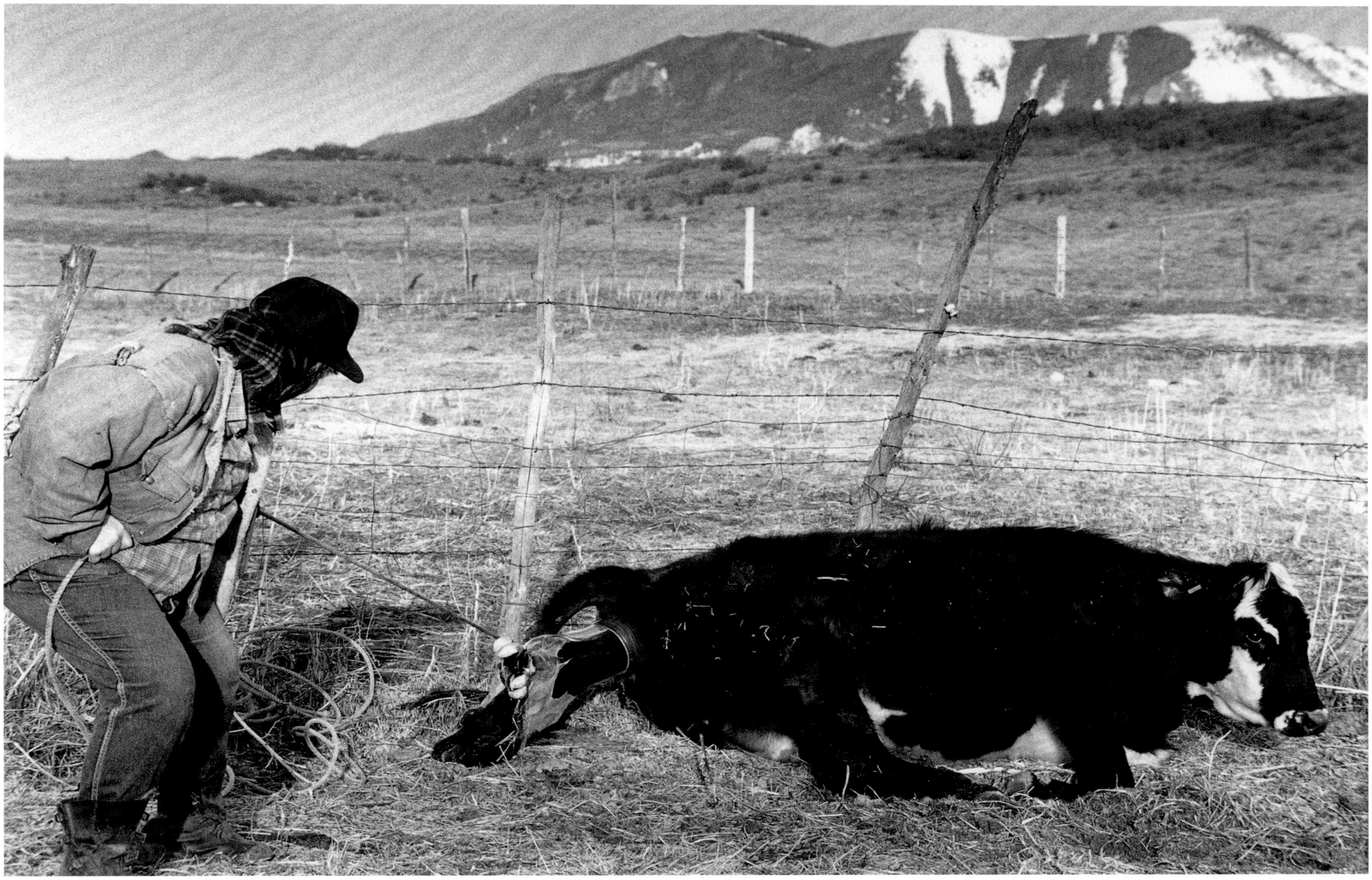

CYNTHIA MAY, PULLING A CALF, S BAR S RANCH
1987

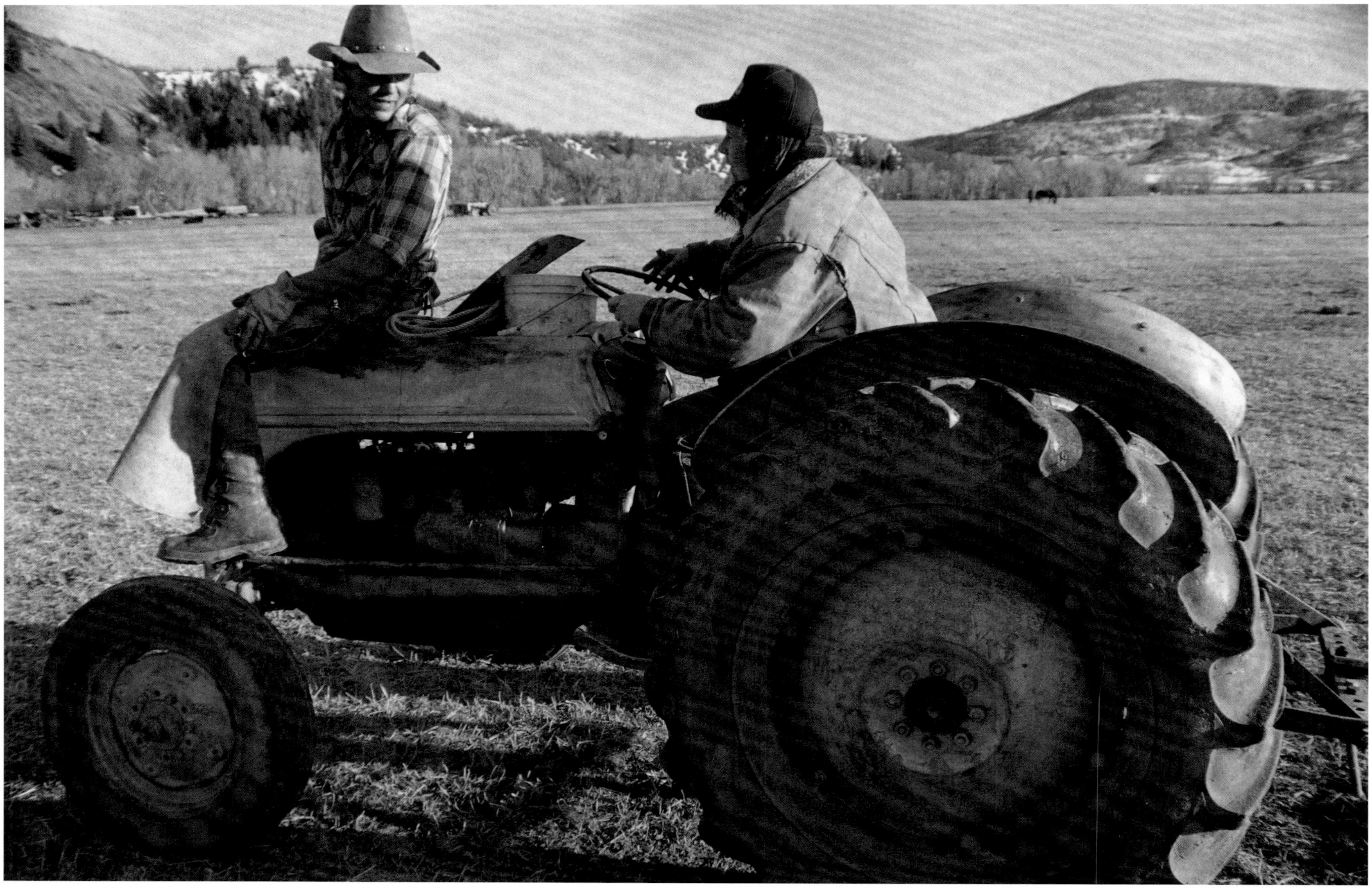

CYNTHIA MAY, S BAR S RANCH
1987

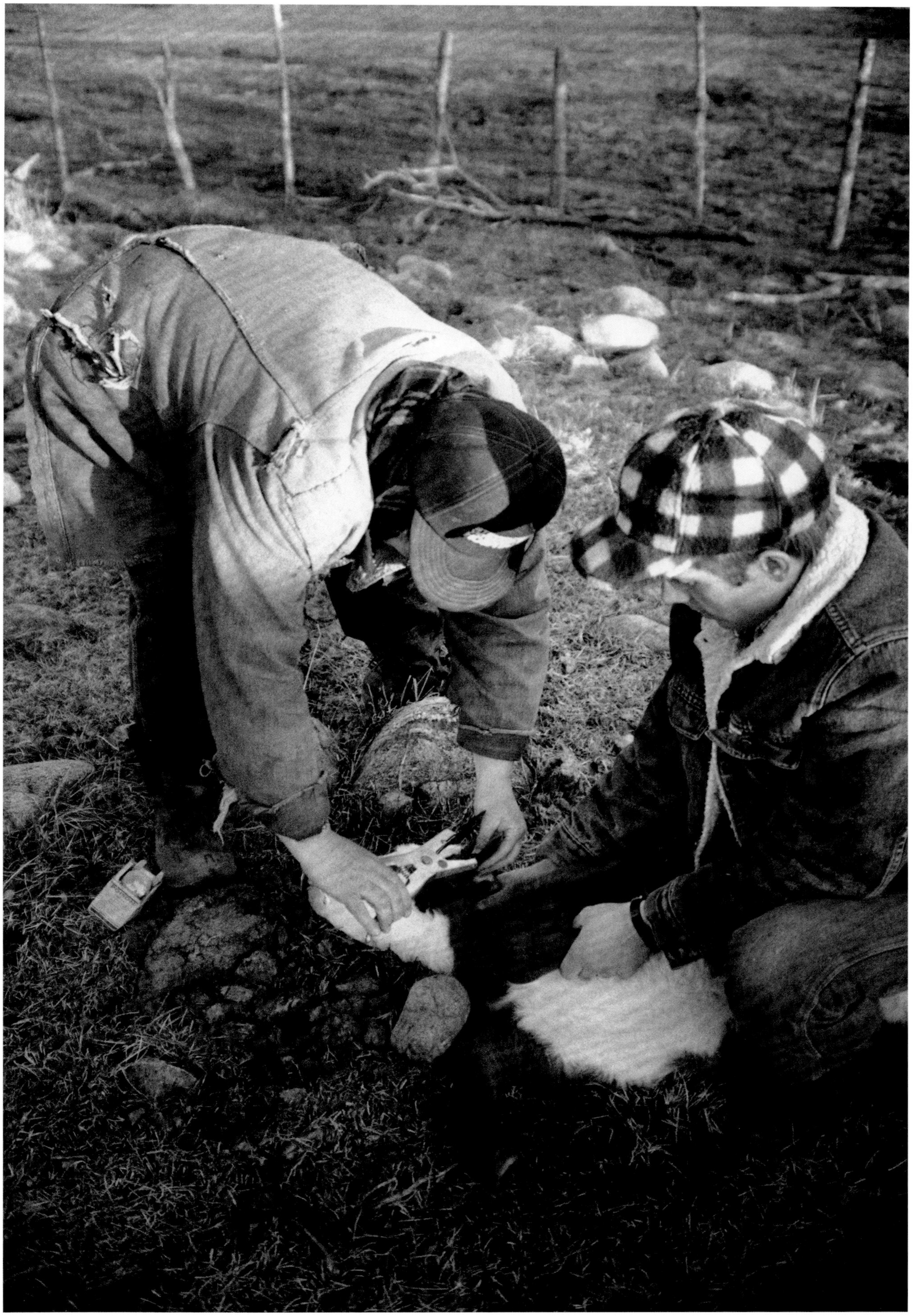

CYNTHIA MAY AND SON SCOTT, EAR TAGGING A CALF, S BAR S RANCH
1987

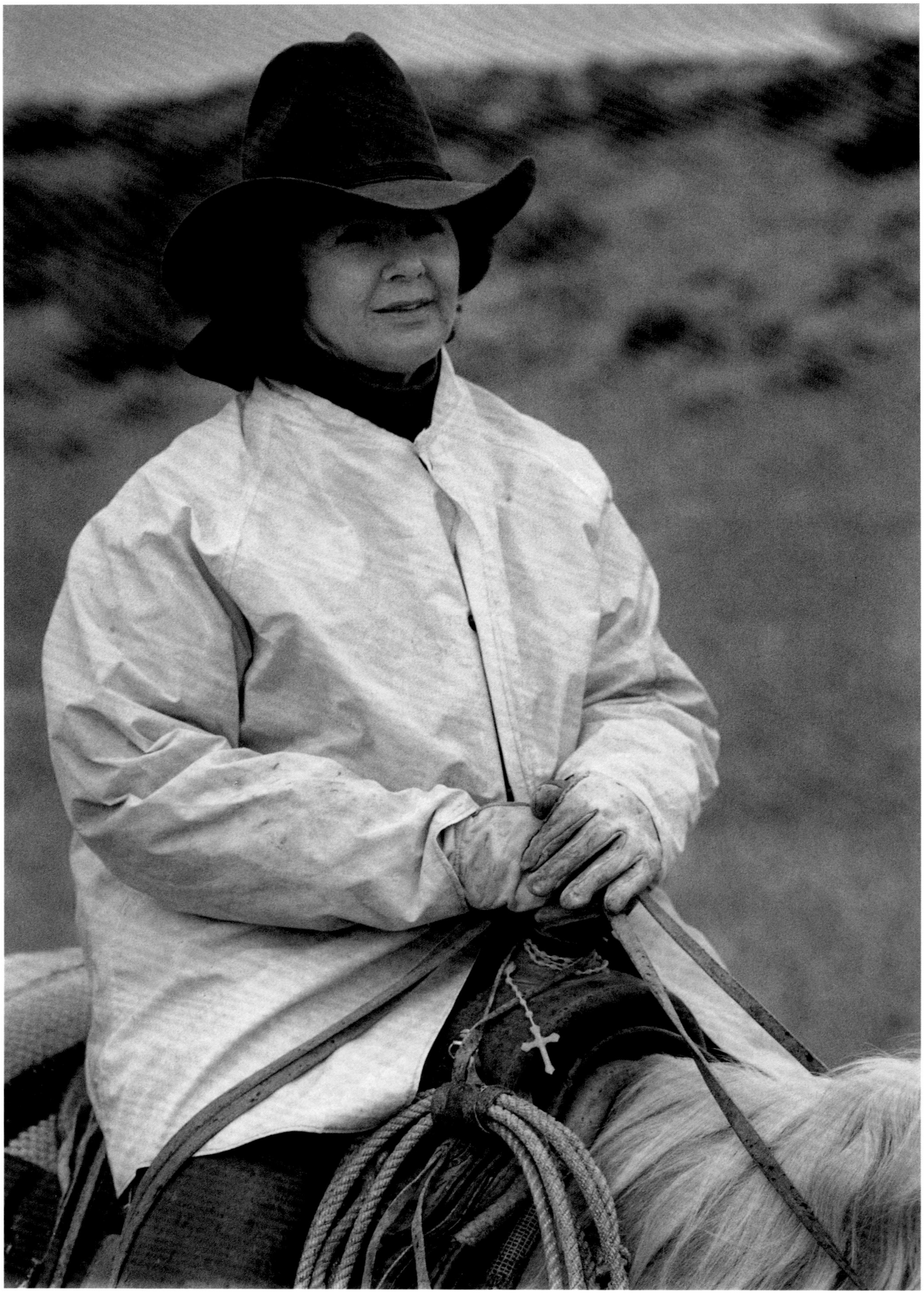

POLLY DICKINSON, VERMILLION RANCH
1987

Dickinson Family

Vermillion Ranch Ltd., Maybell, Colorado

"The back of beyond" is a good way to describe where Polly and Dick Dickinson's ranch lies. It straddles the extreme southwestern part of Wyoming at the corners of Colorado and Utah, not far from the notoriously unruly Rock Springs, Wyoming, where Polly was born and raised.

A lithe, attractive woman of fifty-eight, Polly Dickinson is a pistol, a strongly opinionated western woman who learned to speak her mind at an early age and is passionate on the subject of family:

The extended family is what is missing in the world today. We had it and I cherish it. Dick's parents were a part of our early married life. His dad was the anchor and the leveller. His mother was a good storyteller. She had been raised in an eastern finishing school. It fascinated me that she could come into this ranch life, put on pants and boots, and get down to mother earth and still grub around. They were wonderful and they were fun! It was great for our kids as they grew up to have that closeness with their grandparents who were well educated and didn't fall into the trap of narrow vision. That kind of trap can happen easily on a ranch.

Polly and Dick met at a county fair. His family raised sheep on one side of the mountain (in Utah); her family raised cattle on the other side. They got together and the end results are a solid marriage, four fine children, and a big, well-run family ranch on which everyone in the family works hard and long hours. Each family member has his or her area of responsibility. Wright, Jr., oldest at thirty-four, is the mechanic; he keeps the equipment running and takes care of all the farming. He is also a county commissioner and travels a great deal with his sister Jeanne on behalf of the ranching industry. Jeanne (thirty-three) does all the buying of bulls for the ranch and is the family bookkeeper. She lives in town with Polly's mother. Dee Dee (thirty-one) is responsible for all of the calving and helps younger brother Marc tend to the cattle. Marc (twenty-six) is responsible primarily for the cattle. Dick and Polly are there to help and give advice.

Polly is an outspoken defender of the ranching way of life and clearly every bit her husband's equal. He affectionately refers to her as "the hub of the outfit."

Polly feels that she and husband Dick have a good marriage:

You have to have the belief [that your marriage will make it] and make that commitment . . . and admit that we irritate the hell out of each other at times! My favorite story is the time Dick and I were going down to the corrals to work cattle with neighbors and our hired men. Dick said to me, "Polly, there's three things I want. I don't want you to say "No" in front of the men; I don't want you to second-guess me; and I want you to do what you're told when you're told." So I said, "Okay, Dick, I'll be obedient and submissive—and frigid!" And he said, "Oh shit!" Oh sure, we do holler and cuss at one another, but we've learned to not take it personally. I can always walk away and go to the house!

In the early years of their marriage, Polly's only real fear was that if Dick were to die and leave her the ranch while the kids were all little it would be too much for her to handle alone:

But nothing happened to Dick and now we have the feed [land] and the stock [cattle] and the kids are big enough and know enough to take it any day. They can either keep it together—they helped put it together—or they can tear it apart. Land and water will tear a family apart quicker than anything else. The greed about those two things will do it. So if they want to tear it apart they can. I will not *become attached to the land. God only gave it to us to use. It is not ours. I could walk away from it today and say I did it some good, so if we have to lose it tomorrow, so be it. We'll go on.*

There are many family ranches that have been torn apart because the father doesn't want to let go and let the kids take over. Not true of the Dickinson outfit:

We always included the kids, early on, in business decisions way, way back. Even when Marc couldn't write his name, the children bought some cattle from a neighbor and that was their first experience of signing the note, borrowing the money, and purchasing the cattle in their own entity. And then it was just a case of we acquire a little land and then we'd buy a few more cows and then increasing the numbers because you have to have a viable number of cows to sustain a family unit—500 cows per family.

When the kids went off to college, they were given an outfit checkbook and they paid all their bills out of that, so they knew that so many cows were paying for that education. It scared the banker to death. But Dick and I thought, "Well, if they are going to blow that money, we might as well find out about it now." It didn't happen and it amazed quite a few people.

One morning Polly, Dick, and I rode out to move cattle. It was a cool day in early April. As we rode along searching for cattle in the thick underbrush along the river bottom, we took up our discussion where we had left off late the night before:

It isn't good enough that a rancher and his family have been here a hundred years and have taken good care of the land and raised food on it. Now we have to defend our position on it. That hurts. The public today doesn't know anything except what they're told on TV or read in the papers or magazines. They hear that one AUM (animal unit per month) costs the rancher $1.61. That is what they think *it really costs us. What it* really costs us *is between $15 and $17 per AUM. We have the cost of our labor, interest on borrowed money, fence repair, road maintenance, water holes, and any kind of land and water development we feel is necessary. The federal government agencies don't do a damn thing except hire people to push paper.*

Now the public has gotten all excited about it and think we've been out here ruining "their" land. They don't realize that since we are out here using that land, that we're putting something back. We're giving them food and wool, giving them a resource that is renewable year after year. If we were ruining it, we wouldn't be here. We couldn't have stayed here a hundred years. That is the bottom line.

If I can just find the goodness in some of these people who hate us ranchers. We find animosity wherever we go. If I say we're from a ranch, oh my god, they think I'm a cattle baron or a land baron. I say, "I'm just trying to raise a family and survive."

We had finished gathering, moving, and sorting the cattle. After lunch we loaded up our horses and headed back to the ranch.

Jeanne Dickinson is an attractive and dynamic young woman, as outspoken as her mother. Her major responsibility on the ranch is buying the bulls to use in the ranch operation, for which she travels extensively to bull sales all around the West. By buying top-grade bulls she is continually upgrading the quality of the cows and their calves that the ranch produces.

The attrition rate with bulls is fairly steady: they get crippled and old, shot in accidents, they disappear or simply die of unknown causes. One day in March, Jeanne and I were clearing a pasture to the east of the ranch headquarters. About 100 head of bulls were going to be moved five miles away to another pasture. A Brangus (part Brahma and part Angus) bull tucked up against some rock cliffs and just didn't want to be moved—period! Jeanne left him and when we came back thirty minutes later after gathering the lower end of the pasture, the bull was dead. She had no idea why he died or what could have happened. The bottom line was that it would cost them anywhere from $1500 to $2000 to replace that bull. That is a big chunk of money for an unbudgeted expense, but such situations are a significant part of the business of ranching.

Jeanne also handles the bookkeeping chores, a necessary drudgery on a ranch. For ranchers who, like the Dickinsons, must obtain lease permits for government land, the paperwork is complex and overwhelming. Still single, Jeanne says the man who wants to share her life will really *have* to share it as a member of her family and the ranch operation. He'll have to be quite a man!

They are isolated out there although many people pass through. Polly tells a story about one time when Jeanne was seventeen and alone on the ranch. The rest of the family was off at various meetings. At one o'clock in the morning she heard a knock at the door. The man standing there, dressed in sweats and tennis shoes with a small pit bull dog beside him, asked Jeanne if she could help with his truck, as it had gone off the road and was stuck. Typical rancher that she was, she took him in a four-wheel drive ranch truck to his stuck outfit but couldn't pull him out. He said, well he had some friends at another ranch and would she take him there. So she did. Later it turned out that he had stolen the truck, was armed and considered dangerous. The law eventually did capture him somewhere in Grand Junction, Colorado.

The whole incident scared Polly when she heard the details. She said to Jeanne, "Why didn't you call us on your truck radio?" Jeanne responded matter of factly, "Well, if he had heard you hollering on the radio, and he sure *would* have been able to hear you, it probably would have agitated him, then I might really have had a problem."

I ask her about cattle rustling:

Oh, I think there has been a little bit of it. About thirteen years ago we hired a range detective— tough guy, he had killed somebody. Once the news got around and people knew that we had hired him, a lot of monkey business stopped. Before that, sometimes we'd find something butchered or find a hind quarter or such. Rustling usually falls into a half percent or one percent of "unknown death loss." Sometimes, though, we never can find a few bulls. They're just gone. We buy good bulls and people know it, so I'm certain that once in a while somebody steals one. So be it. By God, they've got themselves a good one!

Dee Dee, the youngest Dickinson daughter, is thirty-one but going on forty. "Mature beyond her years" is a more graceful description. Dee Dee is quiet and reserved, but once she gets to know you and warms up to you she can be downright talkative about anything that really interests her.

She is responsible for calving out all the heifers on the ranch. She works with a crew of Chilean men that handles the night shift but who must call her on the intercom the moment they bring in a heifer who is having a problem or when a difficult situation with a heifer comes up that they know she must hear about. As a result she never gets very much sleep during calving season. "I'm responsible for the calving because I enjoy it and I figure nobody can do it better than I do because I really give a damn about it—I care!"

Proud of her ancestors and family, she explains,

We're fourth generation. Most families that we've known have eventually split up. This ranch was split up at one time between my grandma and my aunt who couldn't get along well enough to run it together, but we want to try to keep it together.

We run a "low management" operation: we work our butts off but we're not bothering the cattle when they're trying to get bred or in the winter when they're trying to eat and get some fat on them. We don't haul them anywhere. We leaves gates open, start them moving, and they just drift on through to where we want them.

She would agree with her mother's statement that "we have a business here, and it's not that really romantic thing you've seen on the TV screen. It's a hard, harsh life, but the short time you have together, if you make the most of it, it's gratifying."

Walker Ranch, Irish Canyon, Colorado

Wanda and Boyd Walker live eighteen miles south of the Dickinson family on the south side of Irish Canyon. The road down to their ranch hangs precariously on the side of a combination cliff and *very* steep hill with a couple of hairpin turns. It is a juicy drive when a good rain or snowmelt turns the dirt road to a gumbo that can suck you over the edge if you aren't careful.

Wanda was born south of Rock Springs, Wyoming, and raised on a ranch there. A friendly, wiry woman possessed of a quiet sense of humor and a remarkable tolerance, Wanda has the reins to the ranch firmly in hand.

They have no telephone but they do have a radio to which they listen for the weather, livestock prices, and news—in that order. She and Boyd have always worked their ranch mainly by themselves, moving the cattle up into the high country even when the weather was bad. Wanda tells about the time, back in '48 or '49, when the snow was so deep the cattle just didn't want to move. So Boyd got a *big* log, hooked it up by both ends and then to his lariat rope, which he dallied around his saddle horn. Wanda sat on the middle of the log to give it more weight and then they took off, his horse dragging the log to clear a path through the deep snow for the cattle. Several times Wanda fell off because of all the bumps the log bounced over, and had to shout for Boyd to stop so she could get back on. "Back then, when we were younger, we thought it was fun," she laughs. "Lots of comical things have happened over the years."

She doesn't have time to take care of a garden because they are always out moving cattle or checking on them or putting salt out for them up on the mountain. "It's a worry about what you're going to get for your calves. There's so much overhead. You lose one calf or two and that's a pretty big loss in your income. We've been losing quite a few to coyotes." Despite the hardships, Wanda remains solidly committed: "I enjoy working with the animals, taking care of them, especially when they're sick. It makes a person feel good to accomplish something for the animals when they're snowed in or sick and down."

Since Boyd passed away, Wanda has had to rely on the hired man to do the heavy work. The Dickinson family keeps an eye out for her and "neighbors," or helps, with the branding and fall cattle work as they have always done. Whenever she can, Polly helps Wanda feed her calving cows. "I hope to be able to stay here," she says. "I don't want to end up in town, but I probably will."

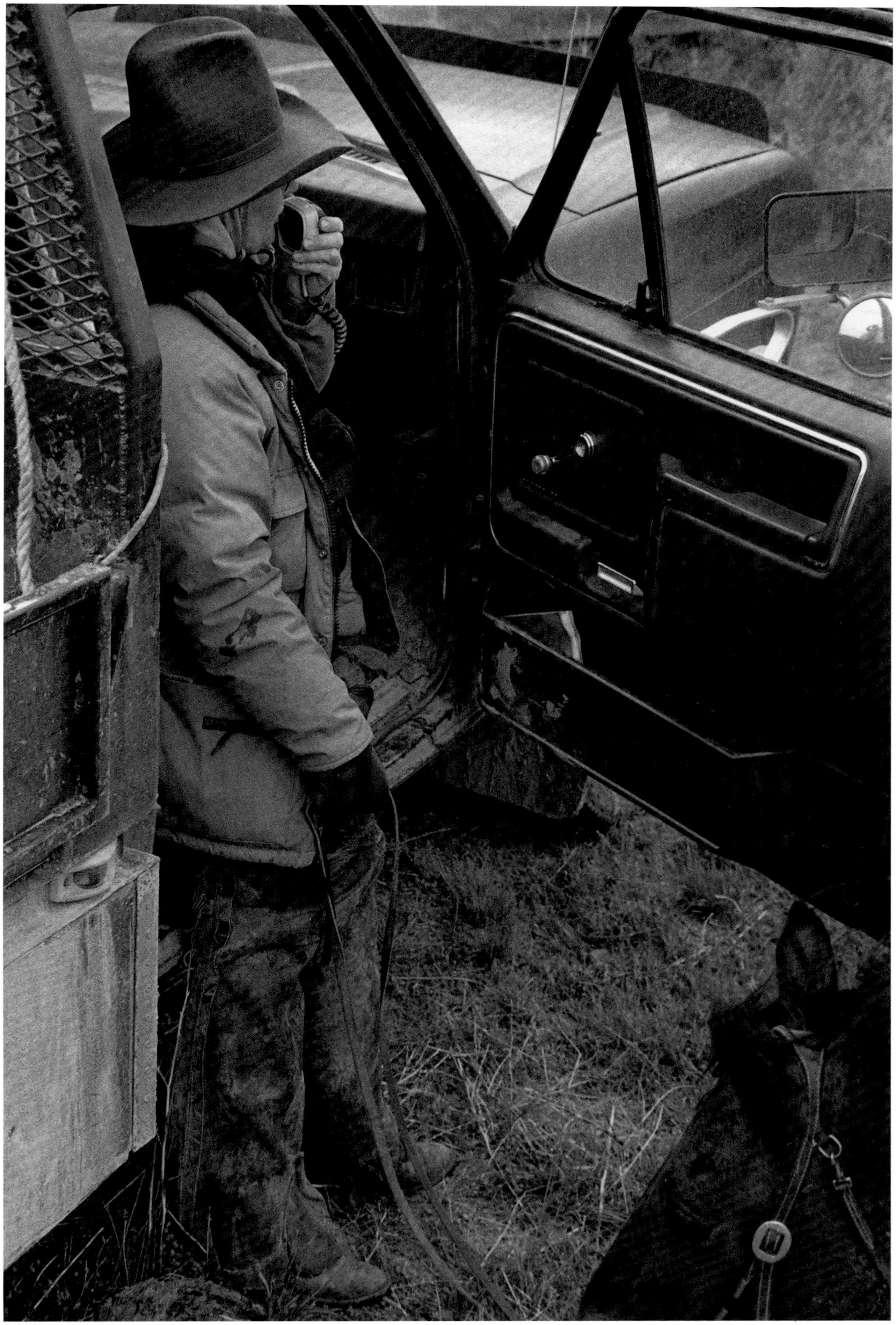

POLLY DICKINSON, VERMILLION RANCH
1987

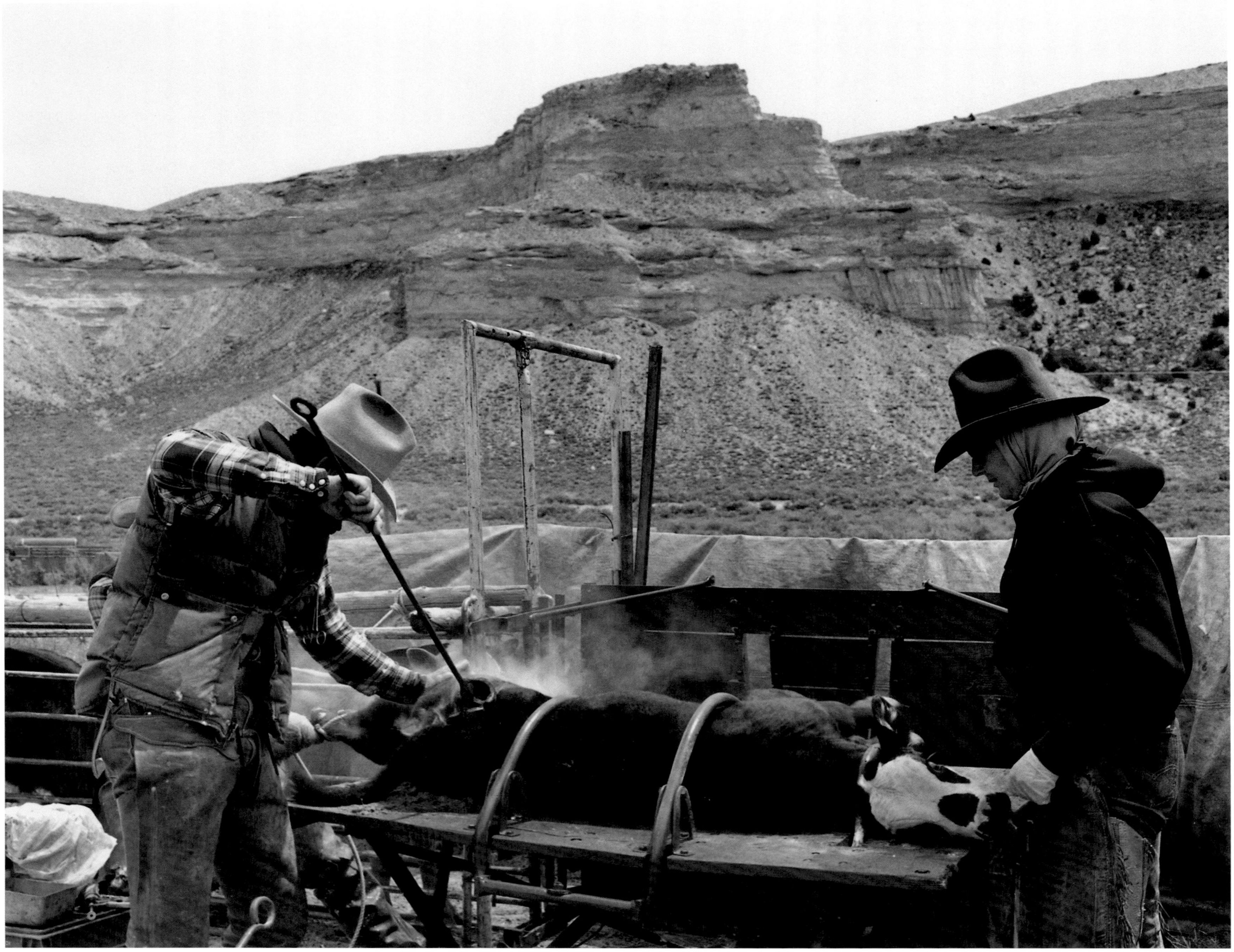

POLLY AND DICK DICKINSON, BRANDING, VERMILLION RANCH
1987

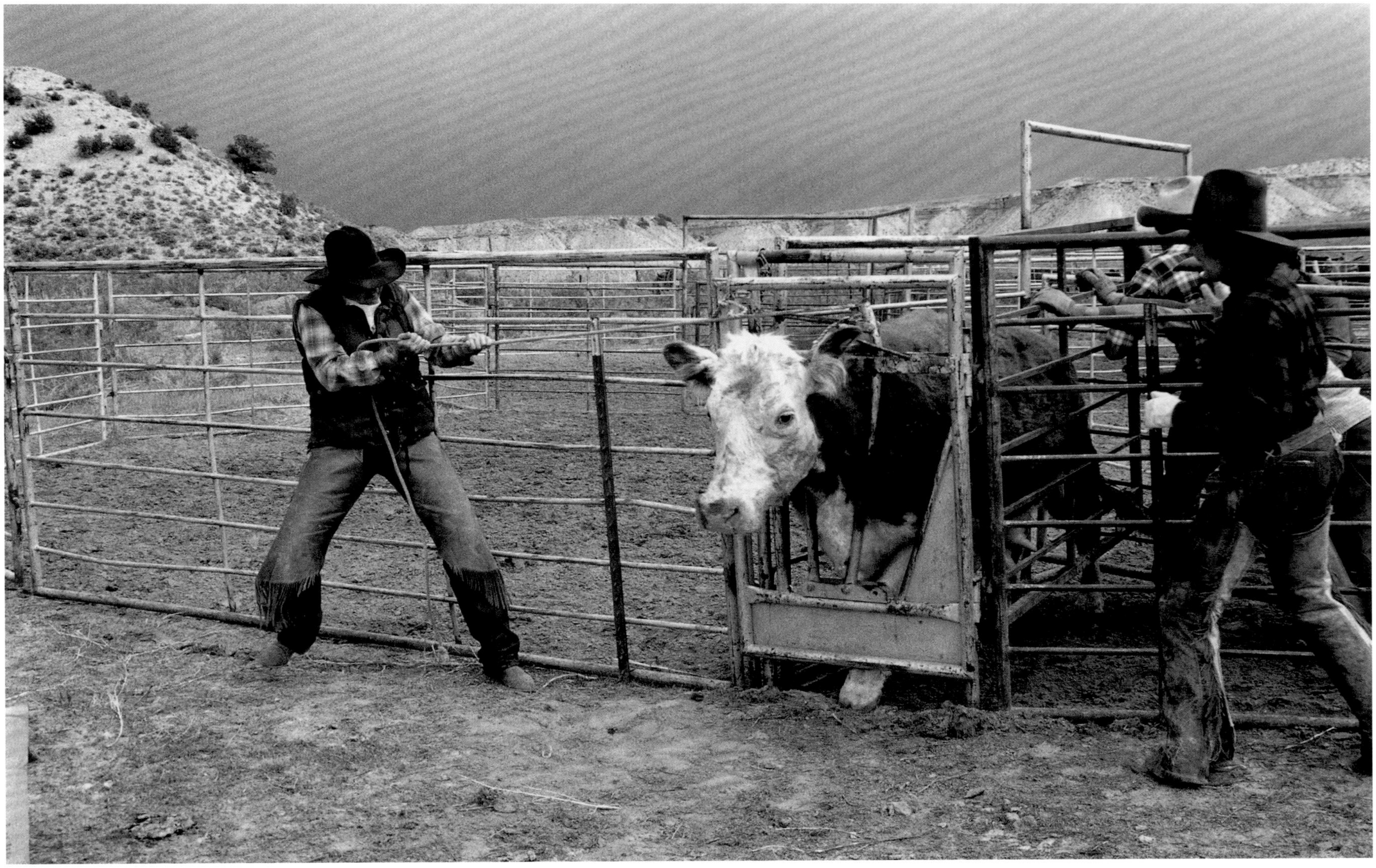

DEE DEE AND POLLY DICKINSON, VERMILLION RANCH
1987

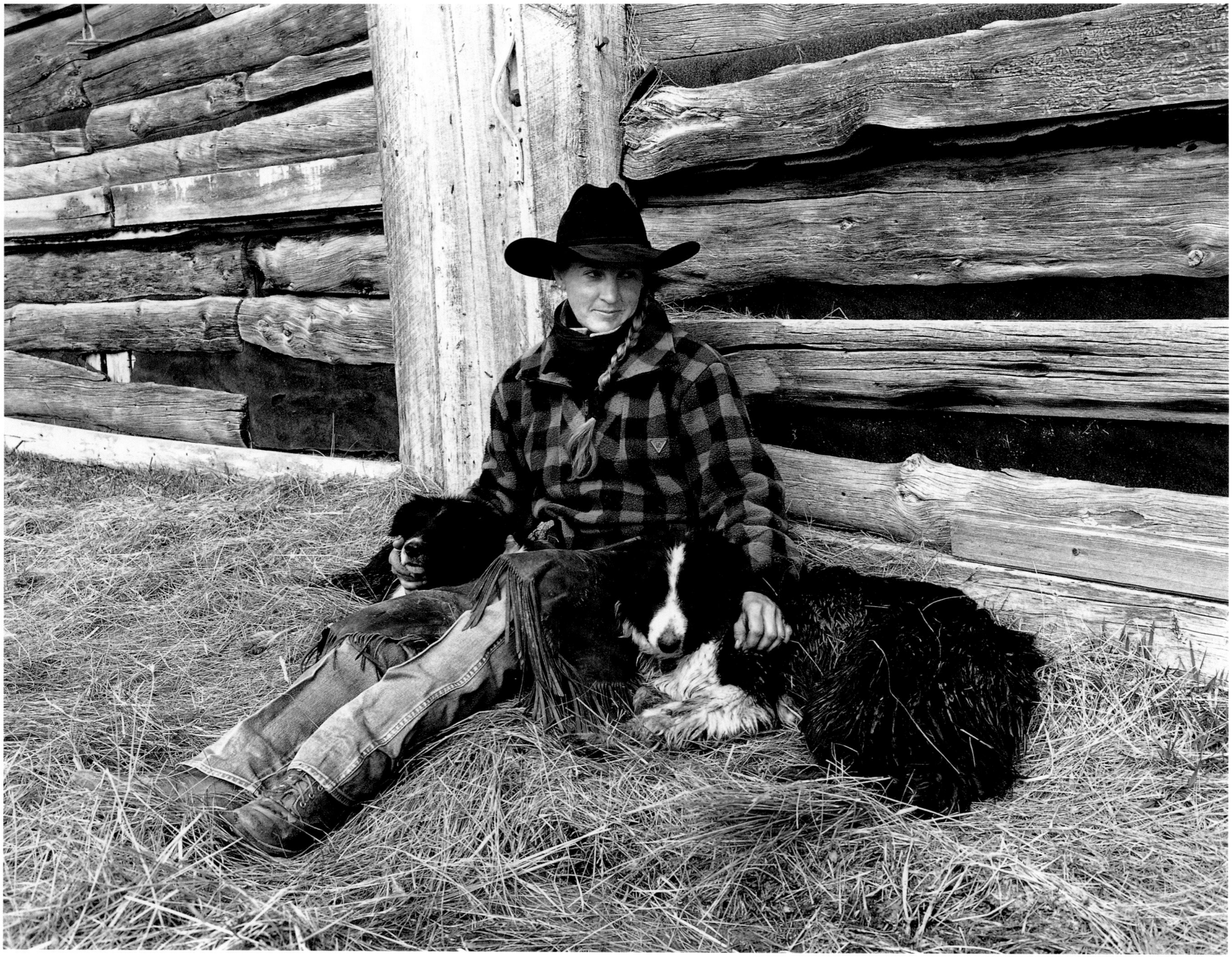

DEE DEE DICKINSON, VERMILLION RANCH
1993

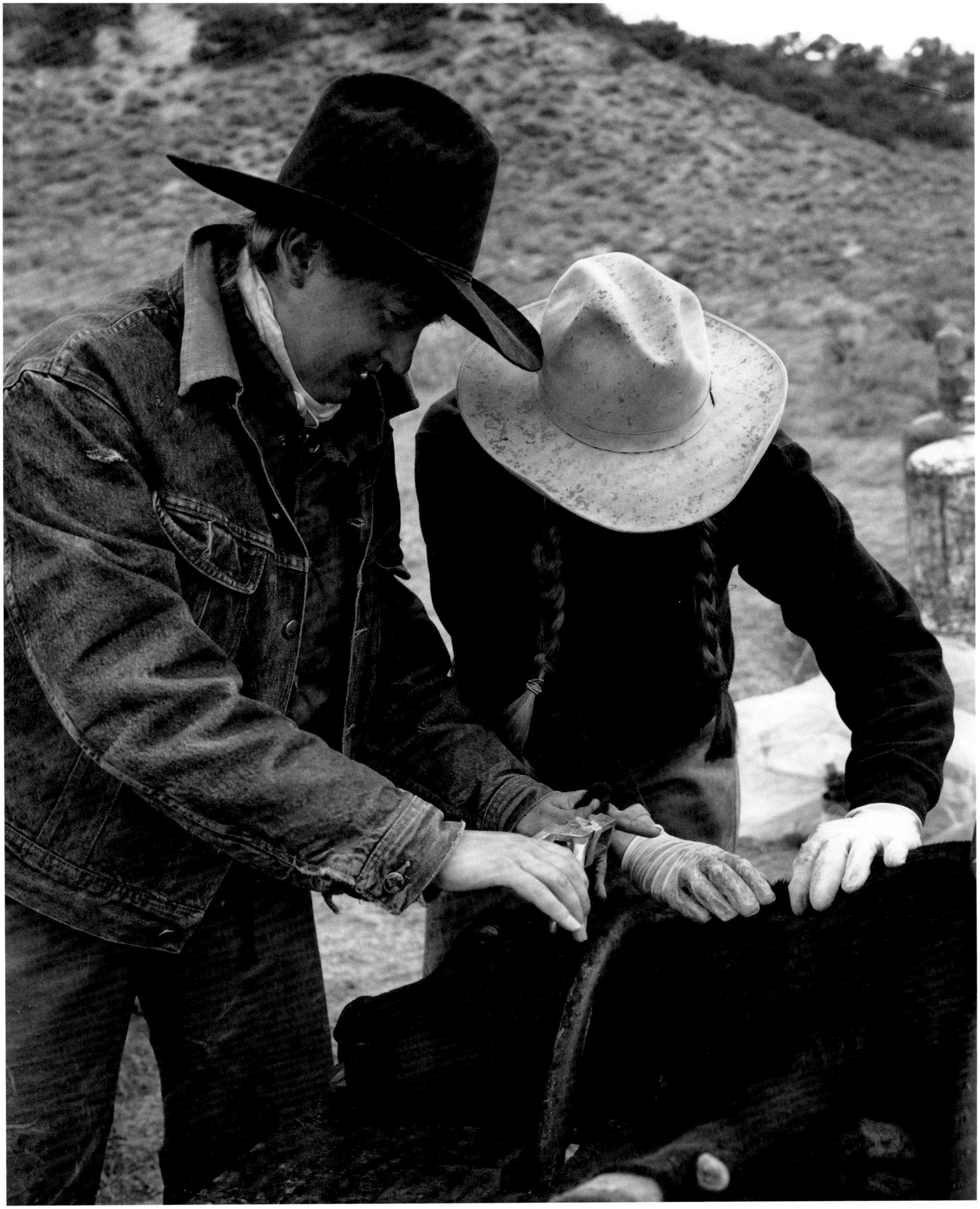

DEE DEE AND JEANNE DICKINSON, VACCINATING, VERMILLION RANCH
1987

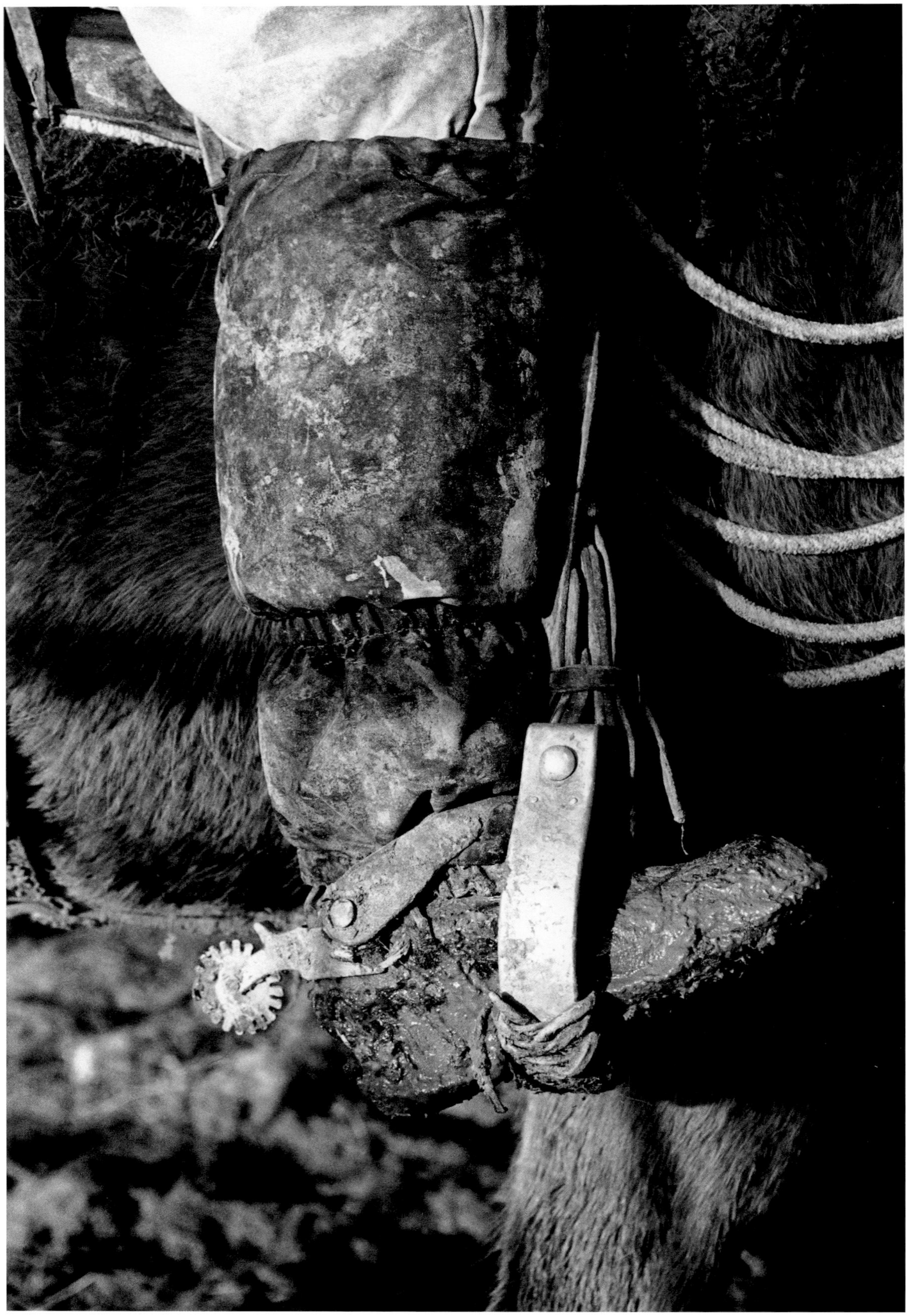

DEE DEE DICKINSON, CALVING LOT FOOT GEAR
1993

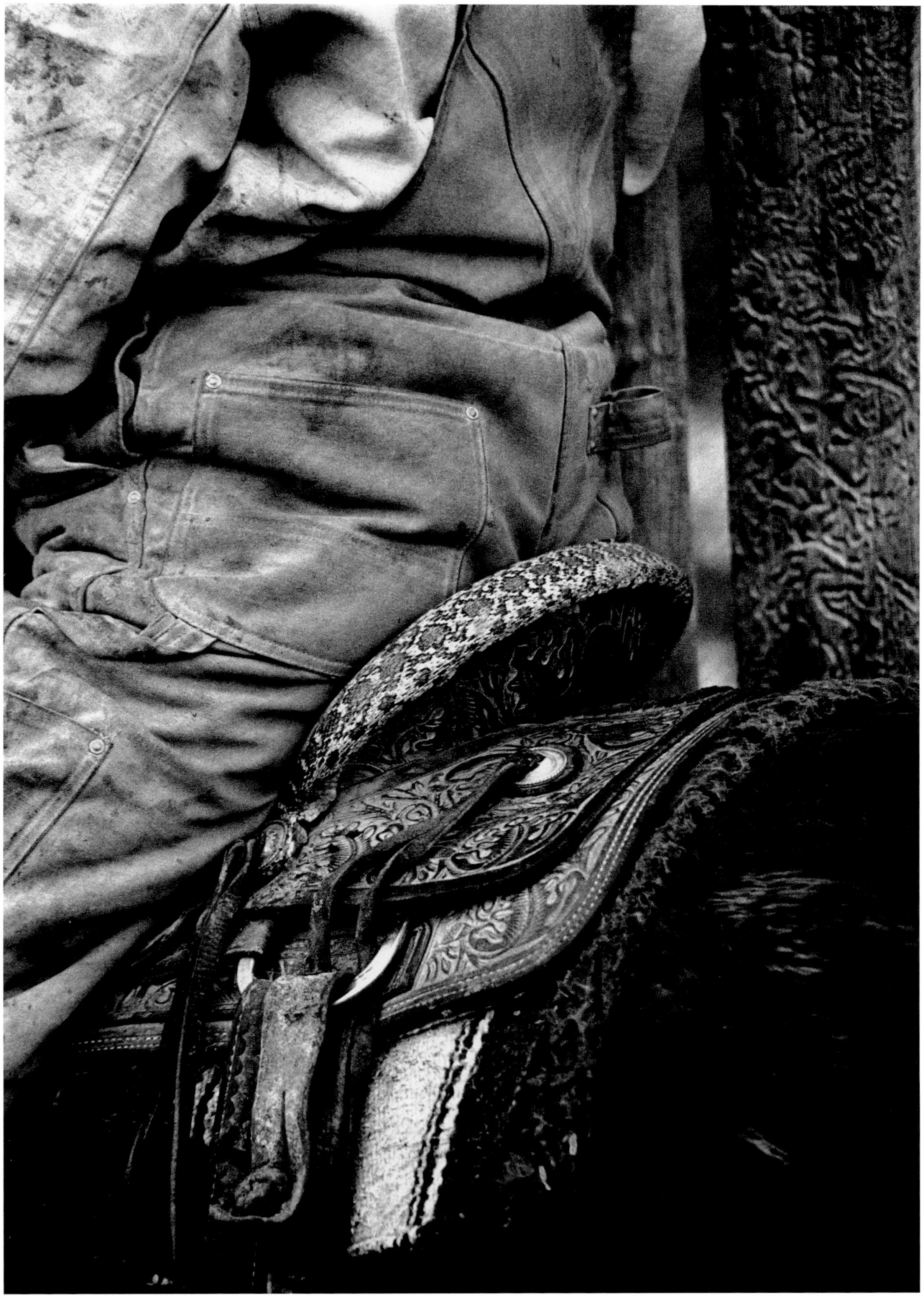

RATTLESNAKE-TRIMMED SADDLE

1993

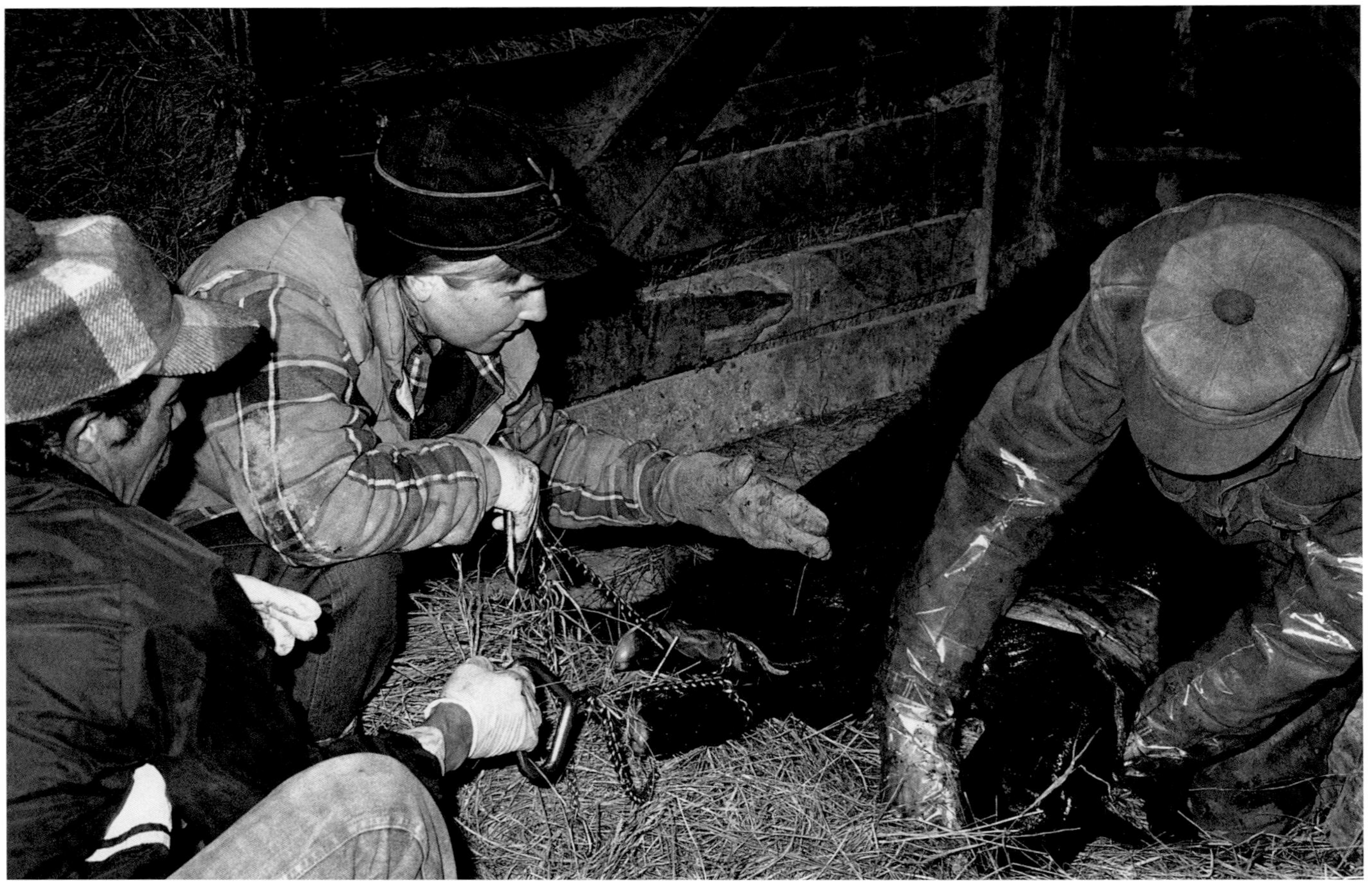

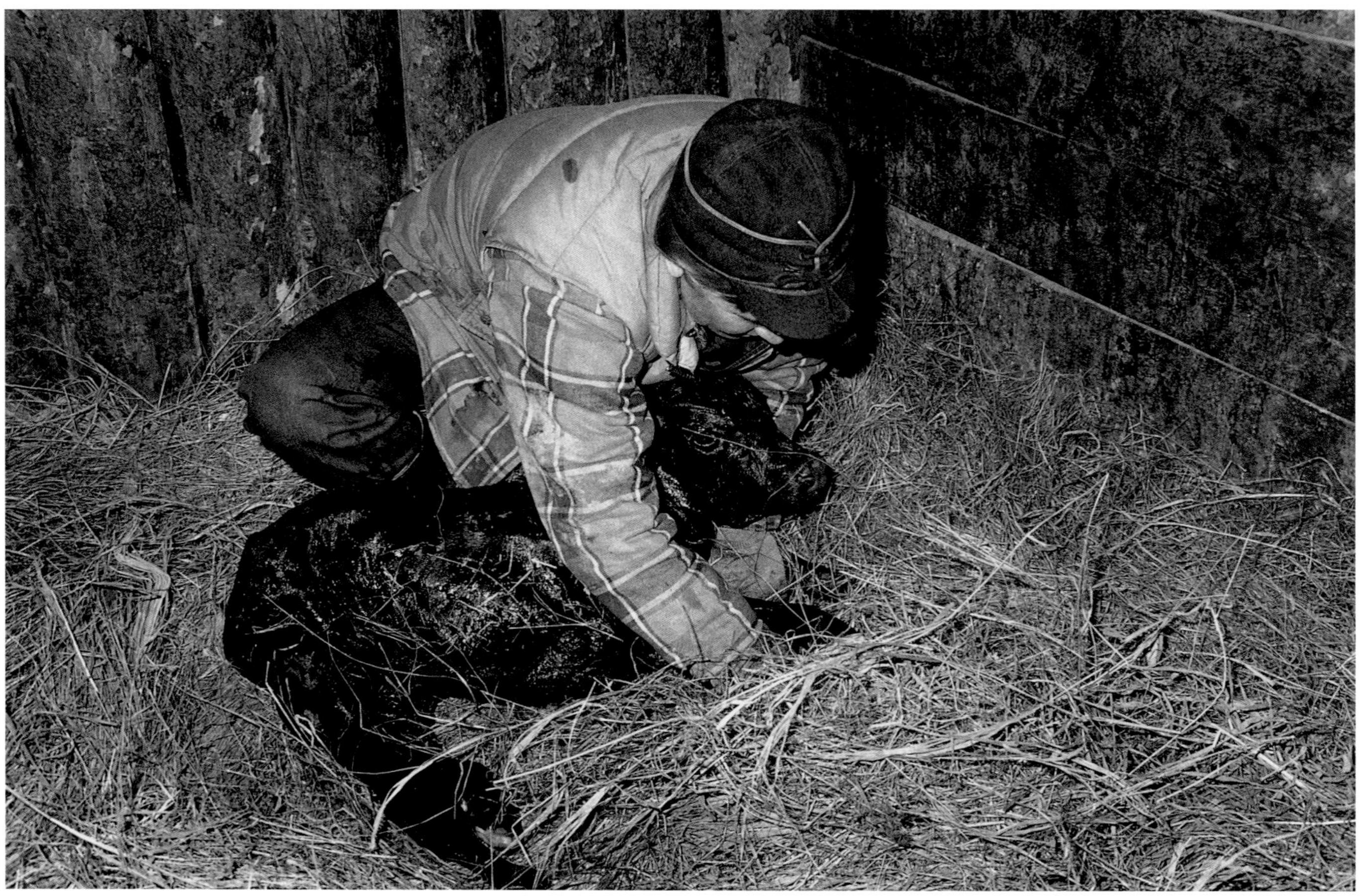

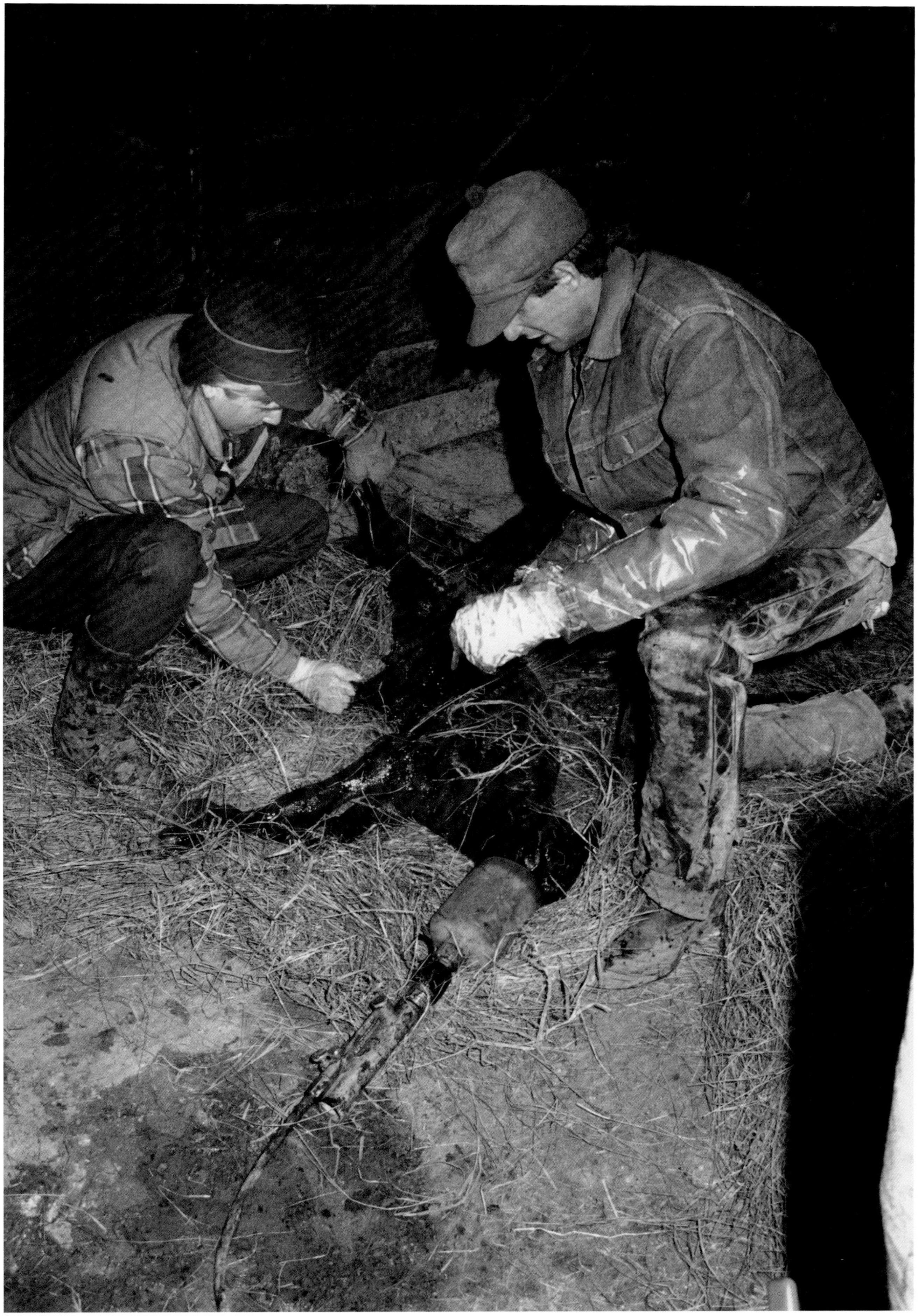

HELPING A HEIFER GIVE BIRTH TO HER FIRST CALF, VERMILLION RANCH
1987

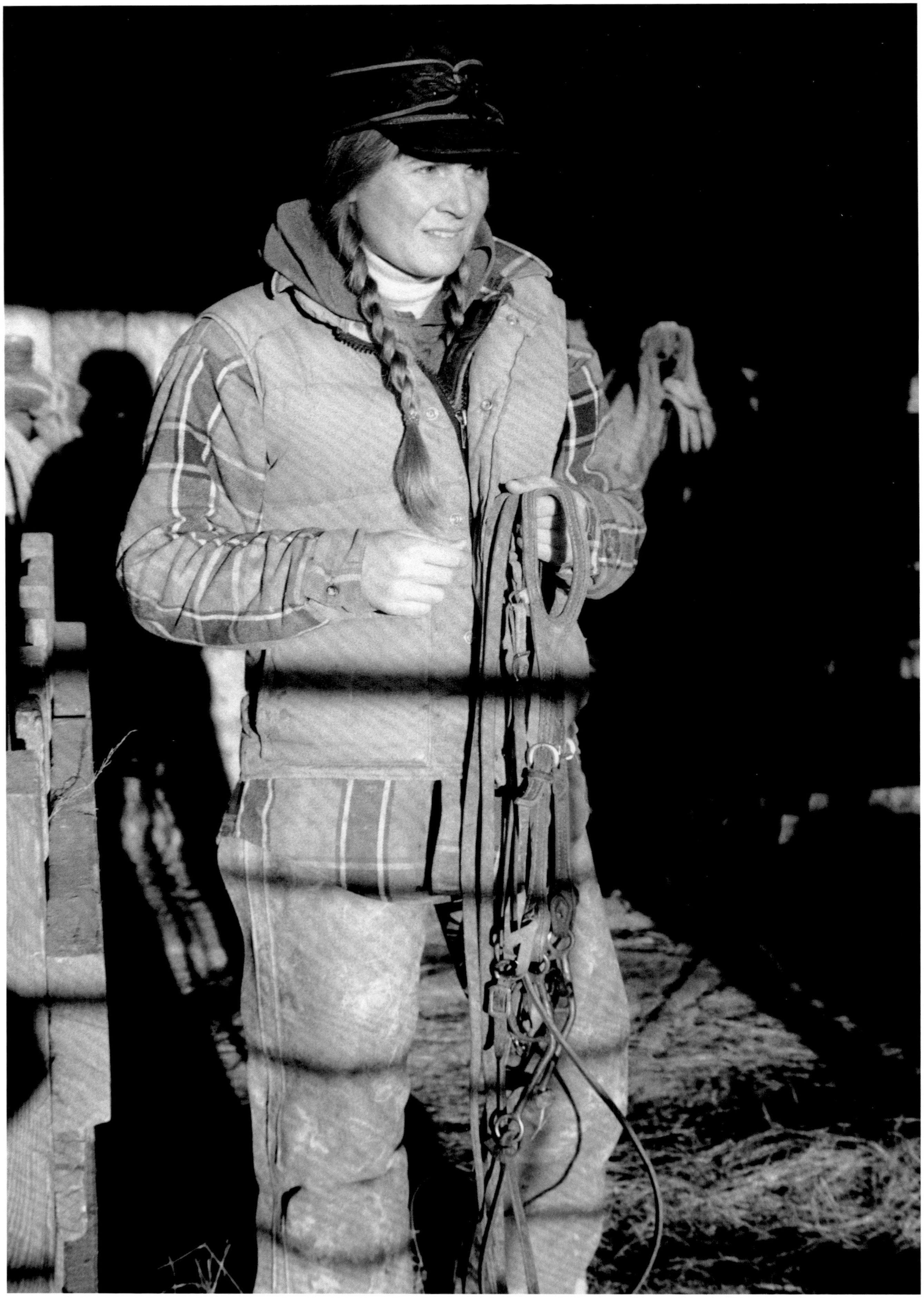

JEANNE DICKINSON, VERMILLION RANCH
1987

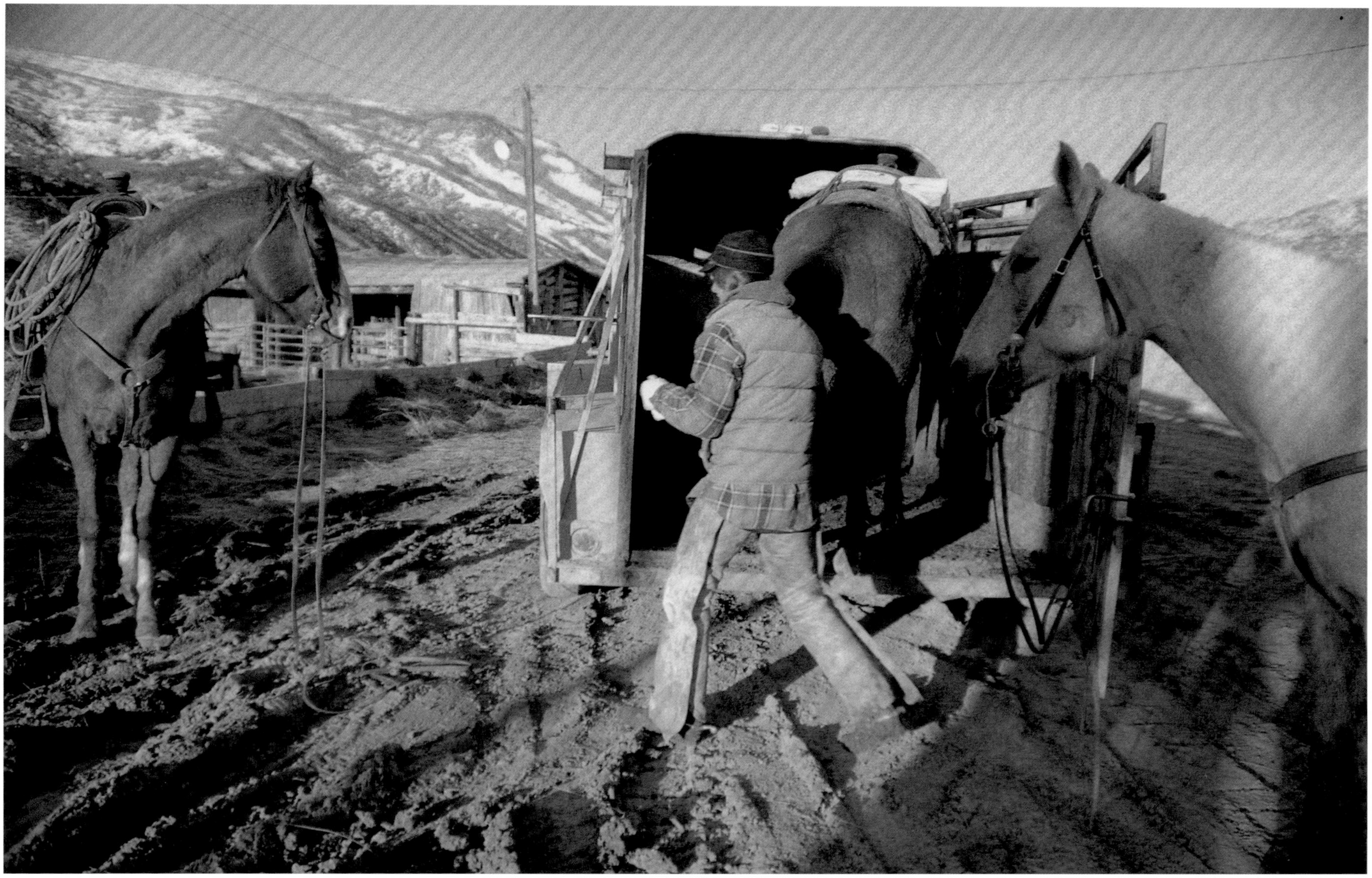

JEANNE DICKINSON, VERMILLION RANCH
1987

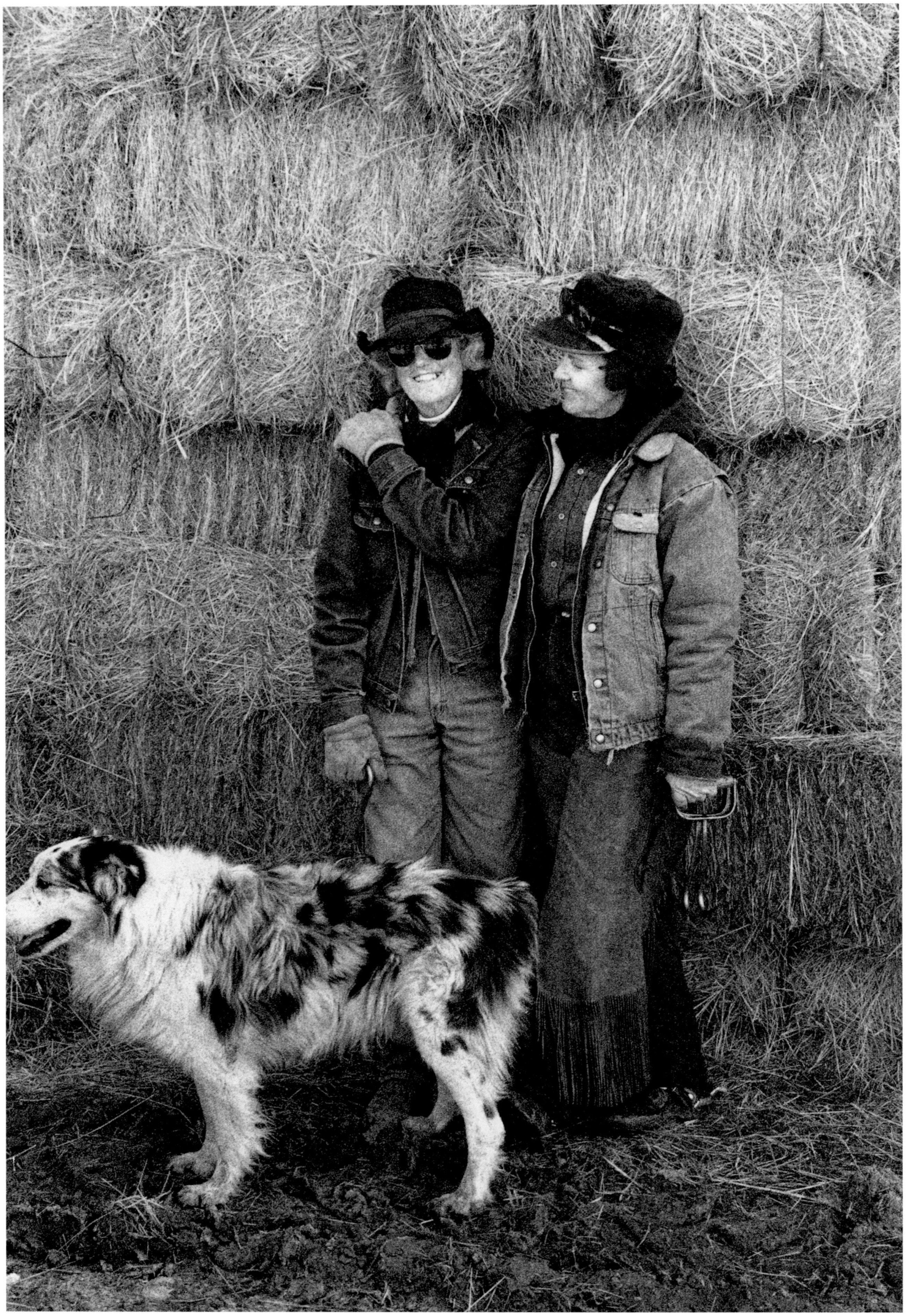

POLLY DICKINSON AND WANDA WALKER, VERMILLION RANCH
1993

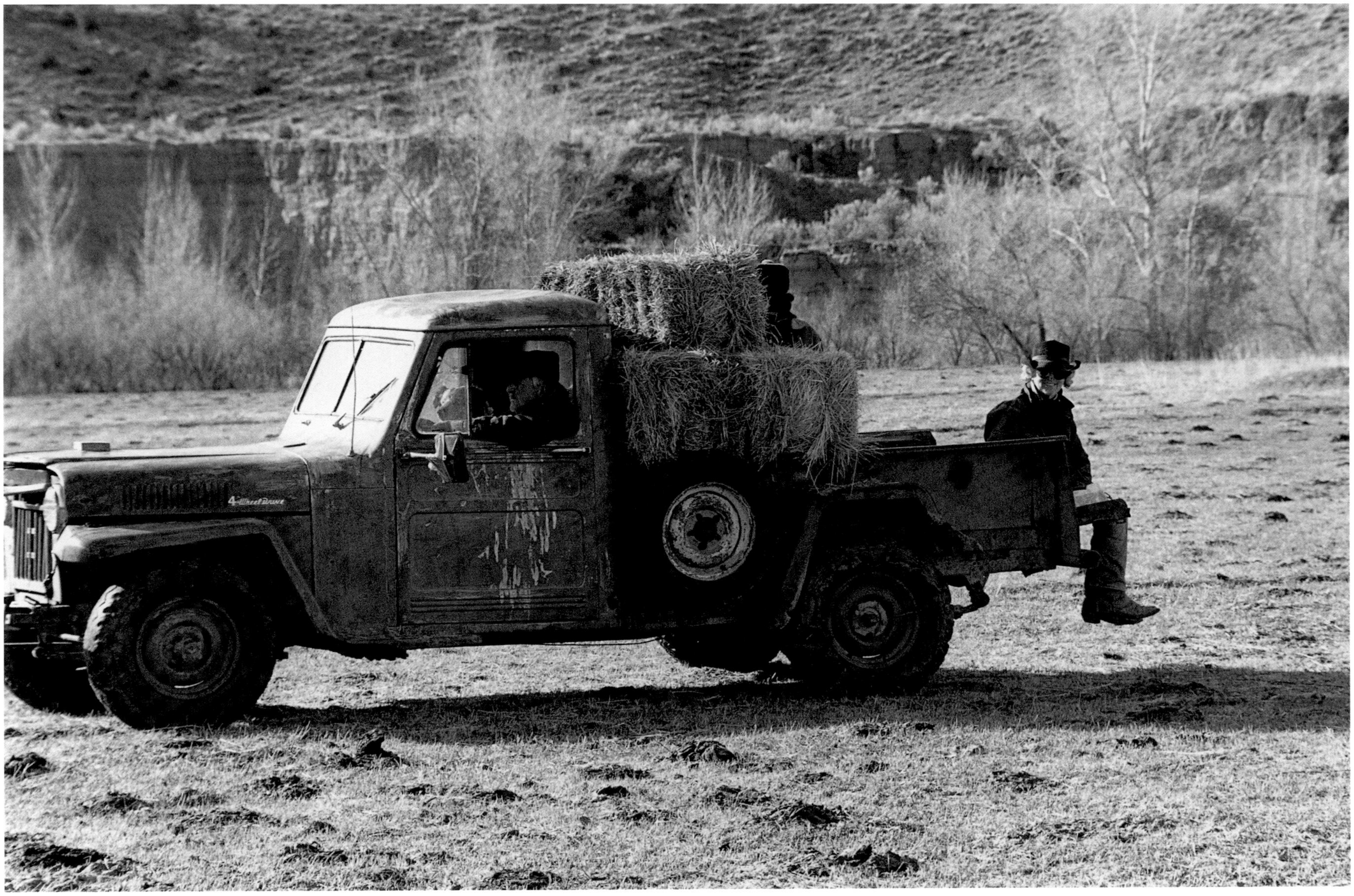

WANDA AND BOYD WALKER, WALKER RANCH
1993

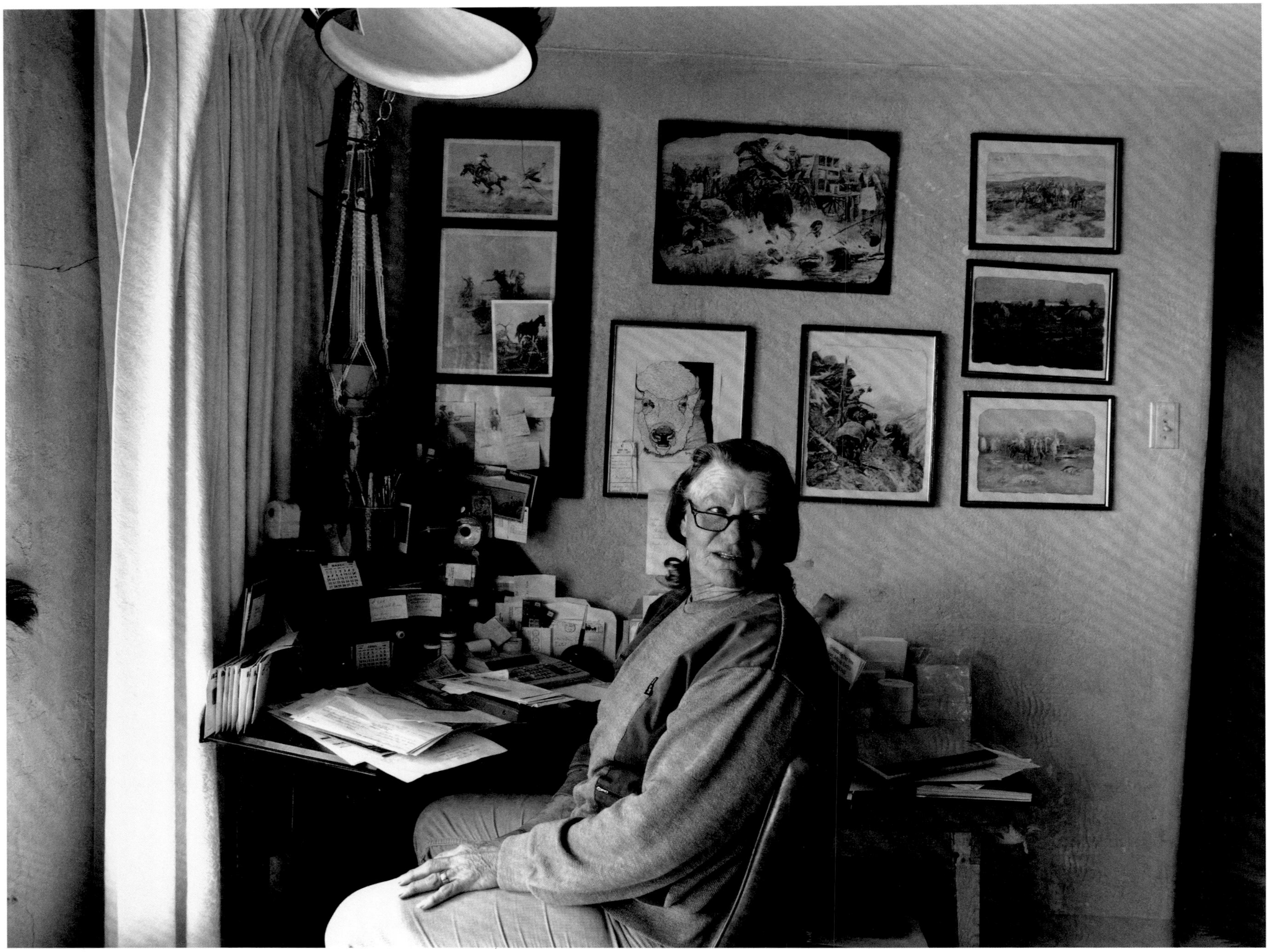

KAREN FARR, FARR CATTLE COMPANY
1988

Karen Farr

Farr Cattle Company, Datil, New Mexico

Underneath a towering pile of thunderheads lies the beautiful wide-open country of the Plains of San Agustin in west-central New Mexico, home to the Farr Cattle Company. It is afternoon on a typical summer day. This is good grass country surrounded by small mountainous hills that are rough and rocky but where water is hard to come by. As a result, the country is studded with windmills. It is also a land of wonderful mirages any time of the year.

Driving up to the Farr ranch headquarters, which sits out in the open with several big old cottonwood trees on the southwest side of the house as a sort of "wind break," I imagined how cold it might get in the winter out here so close to the Continental Divide. I knocked on the door and Karen, a tall, well-proportioned woman with sandy blonde hair done up in thick braids, warmly welcomed me into the sprawling ranch house. Although reserved and quiet, Karen opened up as we had coffee. Several hours later her husband, Dave, came in and, after introductions, began to tease her gently. It is obvious they have a good marriage.

His grandparents homesteaded in New Mexico a little before 1904. Dave's dad, George, came from Corizzo (a town no longer on the state map) to Magdalena, then to the home place here. Their daughter's son is the fifth generation in New Mexico. Although her father had been out here after World War I and fallen in love with the area, Karen was born in Minnesota. Shortly thereafter the entire family moved back out to live in the Four Corners area while he was in the Indian Service. With World War II he joined the army and moved the family to Albuquerque.

Karen recalls their early years:

When we moved to this ranch, we had three rooms and a path—for years—then finally we started to build onto the house. We didn't have electricity. We had a wind charger to charge up our storage batteries, but the refrigerator was propane. It had to work all *the time.*

Our kids, Roy and Amy, went to school in a small town twenty-eight miles from here. We'd drive them the three miles to the highway where they'd catch the bus. When they got big enough to reach the pedals of the truck, they'd drive themselves to the highway, leave the truck and catch the bus. For high school, they went to Reserve, forty miles away. They could hitch a ride with the mail carrier, but they were tied to his schedule so they couldn't participate in sports or extracurricular activities. After that first year we had them board in town with someone and come home to the ranch on weekends.

The Farrs are friendly and open-hearted people. Like most ranchers, they want to keep their ranch together to pass on down to their children, but sometimes situations can make that hard to do.

We had a very tough situation when Dave's brother Ed died [they were partners in the ranch] and Ed's widow demanded his half of the ranch in cash or the land right after his death. We wanted desperately to keep the ranch together, so we had to get a bank loan and hustle to pay her for his half of the ranch. It took a lot of money to do that, so times have been very tight. So it has helped us a lot for me to be working more and more on the ranch. It has saved us a bunch of money because we don't have to hire another man to do the work I'm doing. And I do the bookkeeping. My biggest worry has been and still is being able to make the payment on the loan to buy back Ed's half of the ranch. If the interest rates go up again, we can't make any headway on paying off the principal unless the cattle market goes up, too.

It is a sizeable ranch ("A girl in the county treasurer's office said we were the biggest taxpayers in Catron County.") comprising about 266 sections, including land leased from the state and from the BLM. Water is scarce in their country, and to do any improvements—dig wells, put in windmills, build water holding tanks, lay pipelines, or build fences—they must have permits from the state or the federal government, depending upon whose land the improvement is on. Some of the land they have leased has no water on it at all and would be useless for grazing unless they pipe water to it, so at considerable cost and a lot of work they do just that.

They put in over fifteen miles of pipe and built a Gunite storage tank in the foothills into which they pump water from the old ranch well up a canyon to the top and down the other side into the storage tank. With that they've been able to bring water to two separate canyons for their cattle. The water (and the salt they put out for the cattle) has attracted wildlife that never was there before.

What is a big problem for the ranchers, and for which they get no credit from environmentalists nor from the state game and fish departments, is the unprecedented expansion of wildlife ("game" to hunters) onto their lands as a result of the salt and water that ranchers have made available. For the Farrs, this is the spread of elk onto their lands both deeded and leased. A single elk will eat as much as a cow and compete with the cattle for feed and water. And no one, not environmentalists, nor hunters, nor the state game and fish department, has offered them any compensation nor expressed appreciation for what they have done and are doing—at their own expense.

Dave wrote the state game department about this situation and said that he thought they ought to give him some hunting permits in return for all that we are doing for the elk. Some guy responded with a nasty letter saying, in effect, "We don't accept bribes." He must be stupid to think that! If we could keep the damn elk out we would, but they just tear down fences the minute you put them up. And the game department won't fix our fences on the land we lease from them unless we can prove beyond a shadow of a doubt that the elk broke them down. None of us on the ranch has the time nor the money to lie around waiting with a camera to photograph the elk bulling their way through, tearing down a fence. The principle *of this thing is the real issue. The elk* were not there *until we brought water and salt up there.*

Karen opines, "Our ancestors had to 'make do.' They had to make everything from scratch. This wasn't a problem, but if I had to have one convenience, I'd take refrigeration."

There's so much dirty work to ranching that it's almost impossible to hire anyone who wants to really work. *Most of these young fellas just want to dress like a cowboy and let their legs hang down [ride horseback]. They don't want to repair windmills, work on pipelines, mend fences, or do any of the dirty work. They just want the glamour of horseback work. Hardly anyone anymore knows* how *to milk a cow, let alone* wants *to milk a cow.*

On the subject of women's contribution to ranch life, Karen Farr has a ready response:

Women bring a great deal to a ranch. For example, it seems to me that when a woman is on a ranch, the man is more content. And women bring compassion to ranching. A lot of men seem to be less sensitive. When a cow is down for a long time, I tell the hired man, "keep her tail out of the shit, clean her up, clean her up!" If you or I were lying in bed, you know how good we'd feel about being kept clean. The men don't see or feel that. They look but they just don't see. *They don't see the details. For example, after we've pulled a calf or worked on a heifer, I always like to wash them off; the men don't see that as necessary. Or if the cow or heifer gets torn during giving birth, the men don't put ointment on her. I tell them to do it. Women* are *empathetic.*

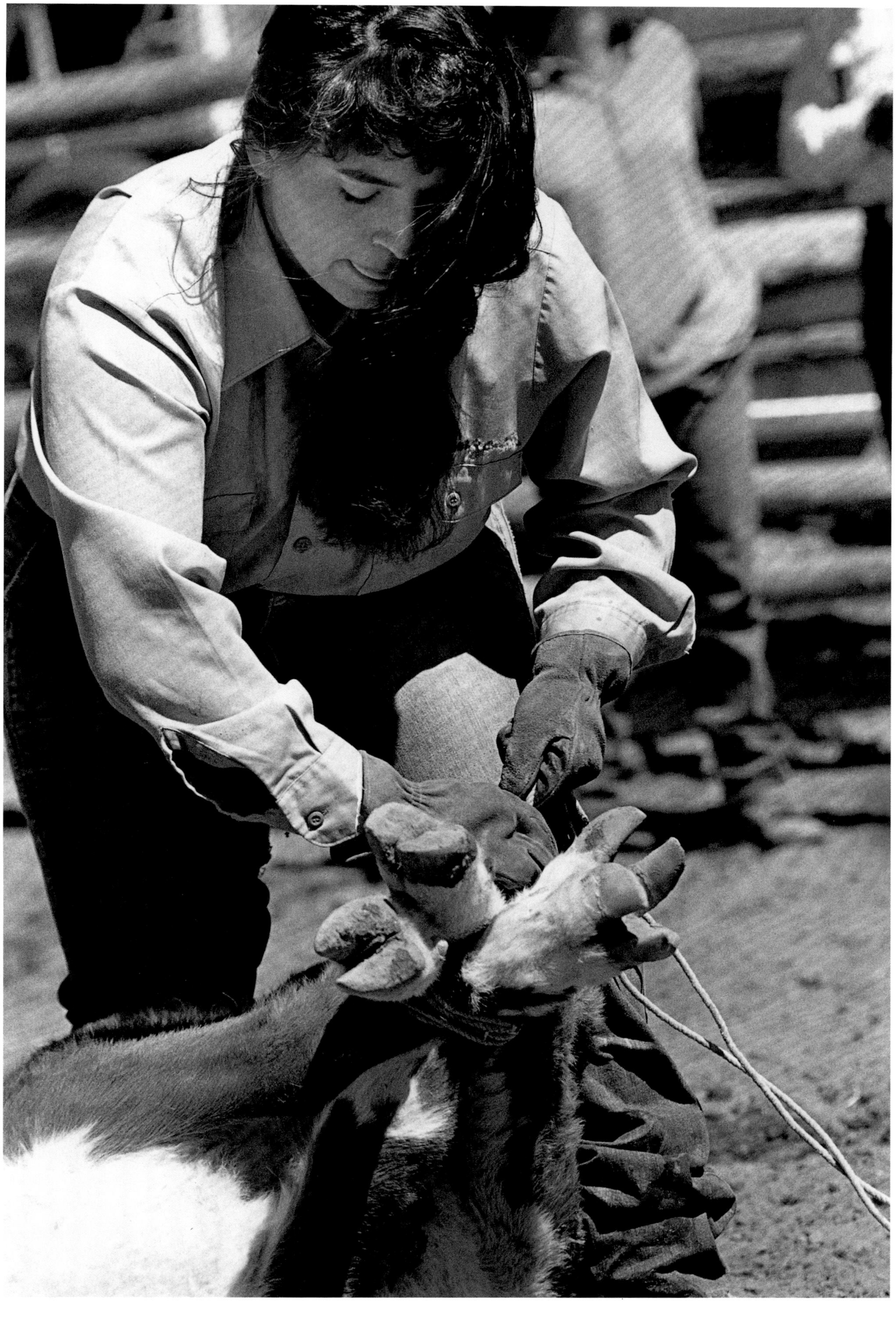

ELENA FARR, SPRING BRANDING, FARR CATTLE COMPANY
1993

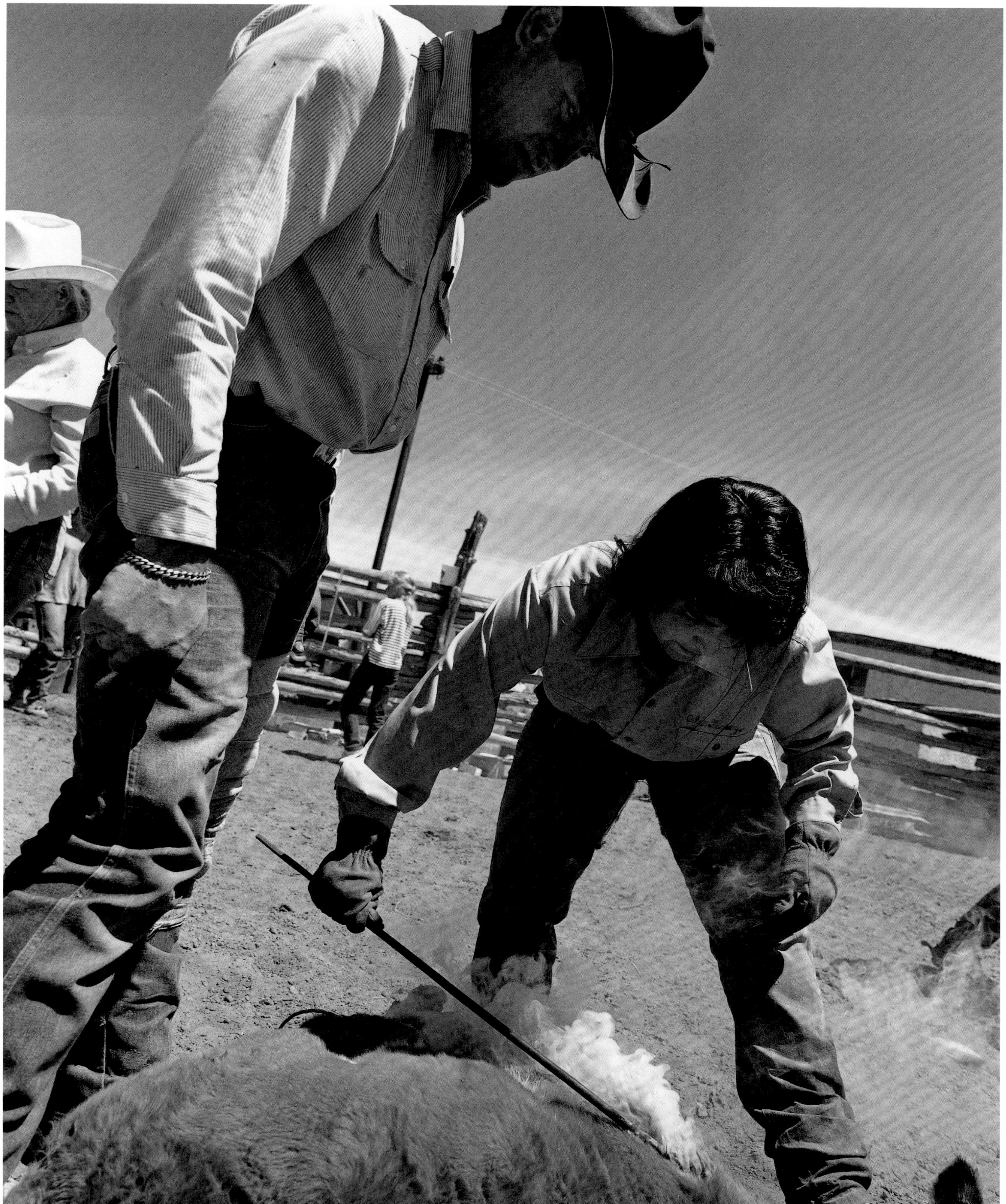

ELENA AND ROY FARR, FARR CATTLE COMPANY
1993

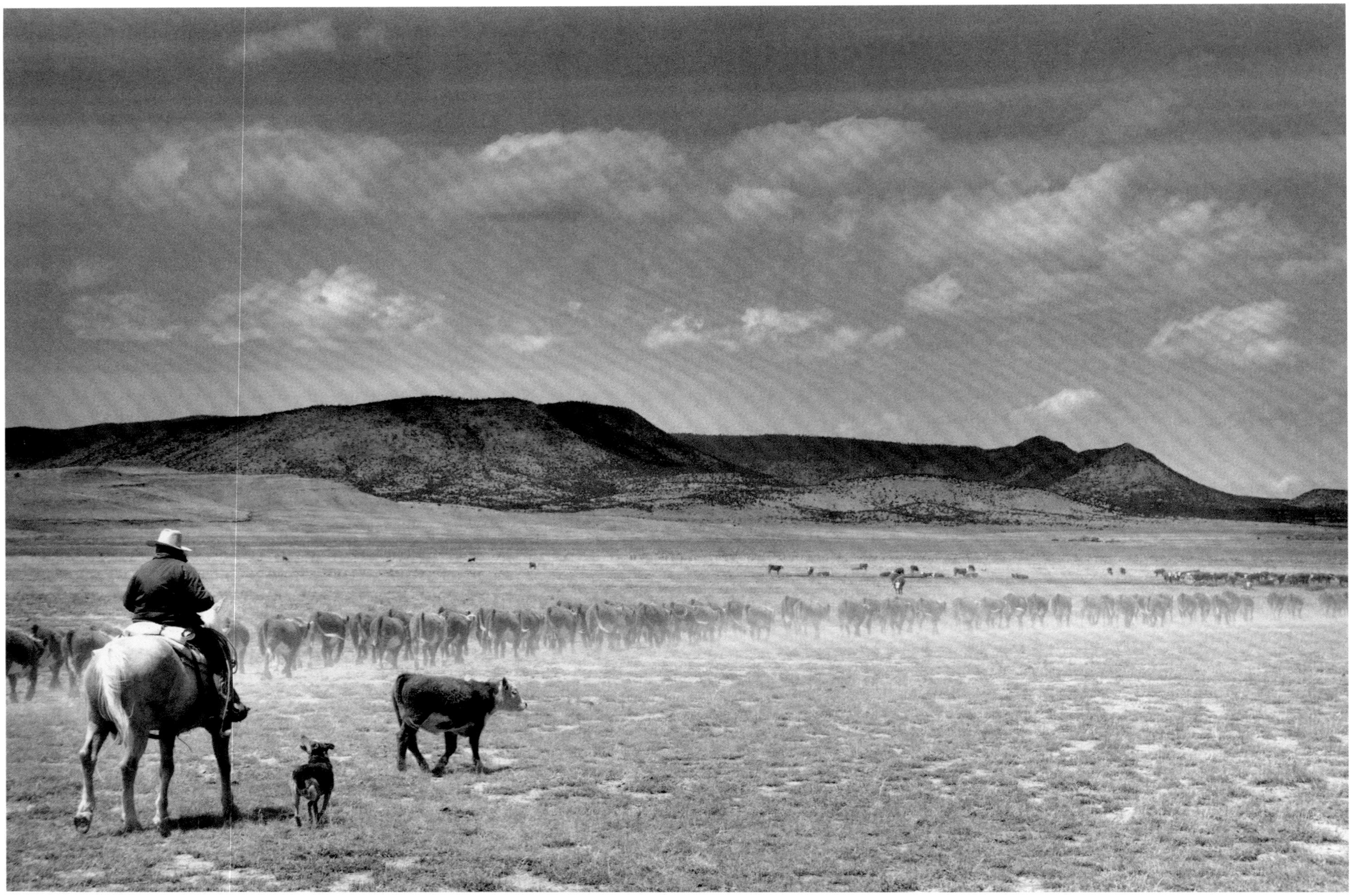

KAREN FARR, FARR CATTLE COMPANY
1988

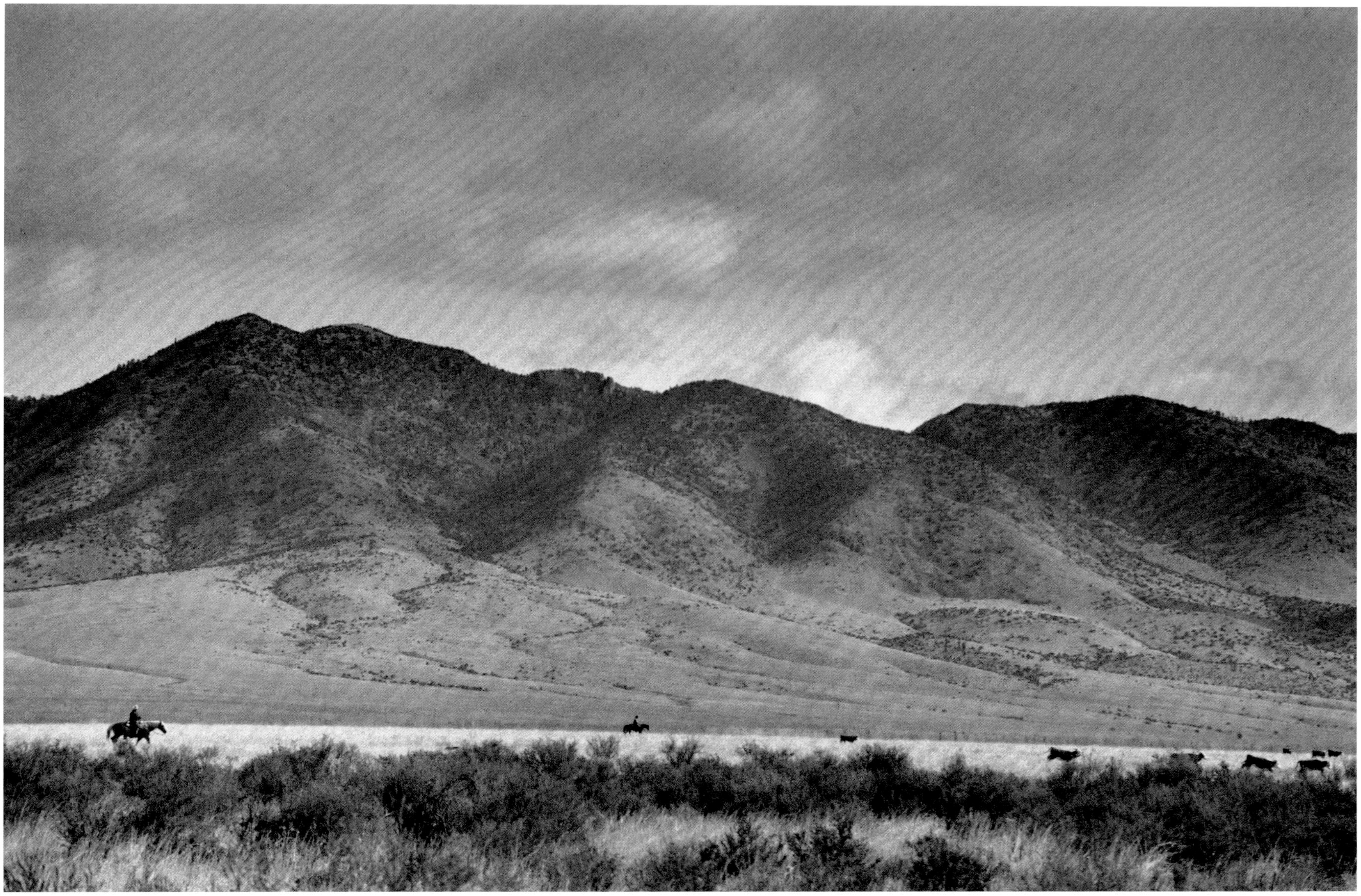

KAREN AND DAVE FARR, MOVING STEERS, FARR CATTLE COMPANY
1988

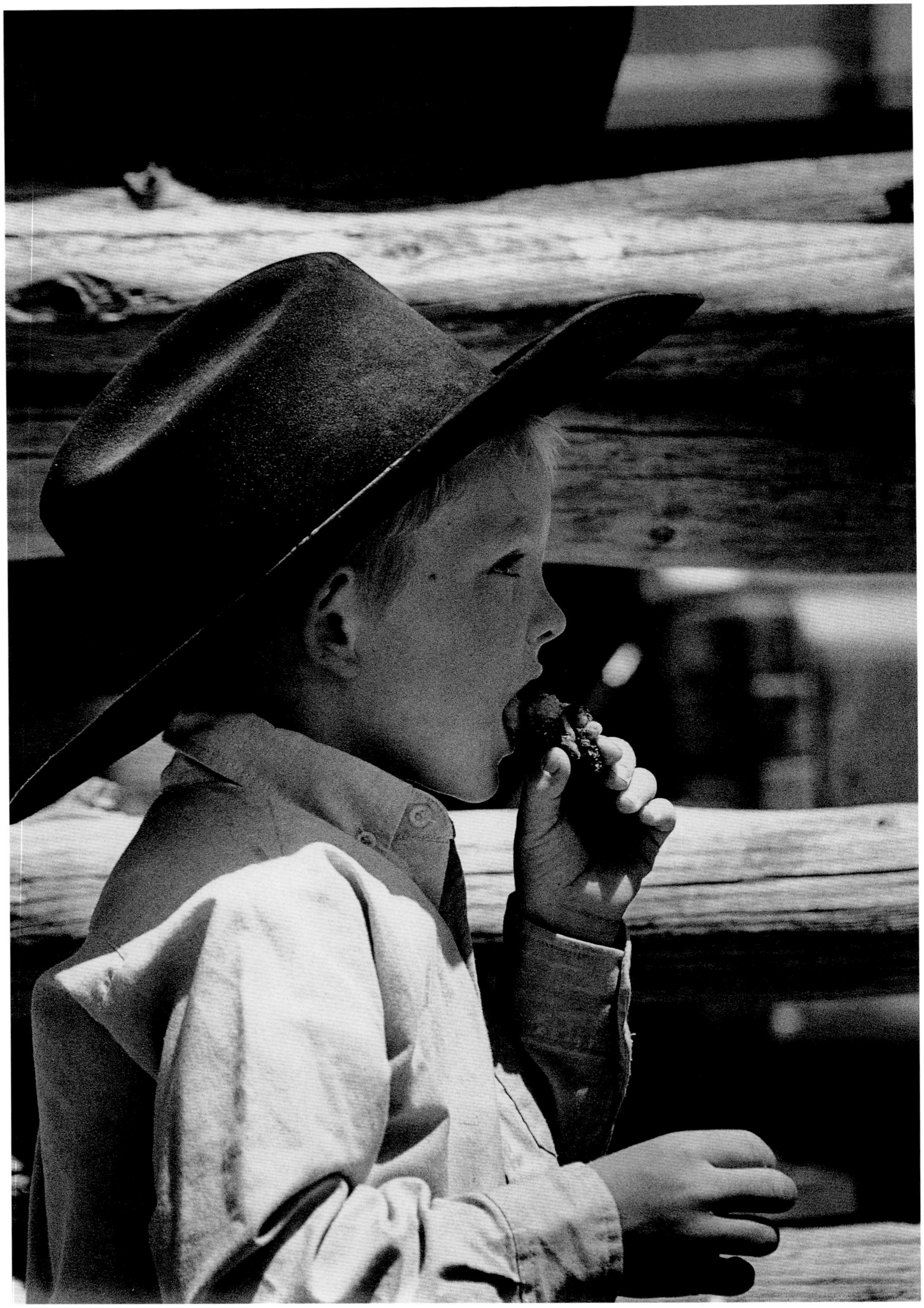

GRANDSON HANK WILLIAMS, FARR CATTLE COMPANY

1993

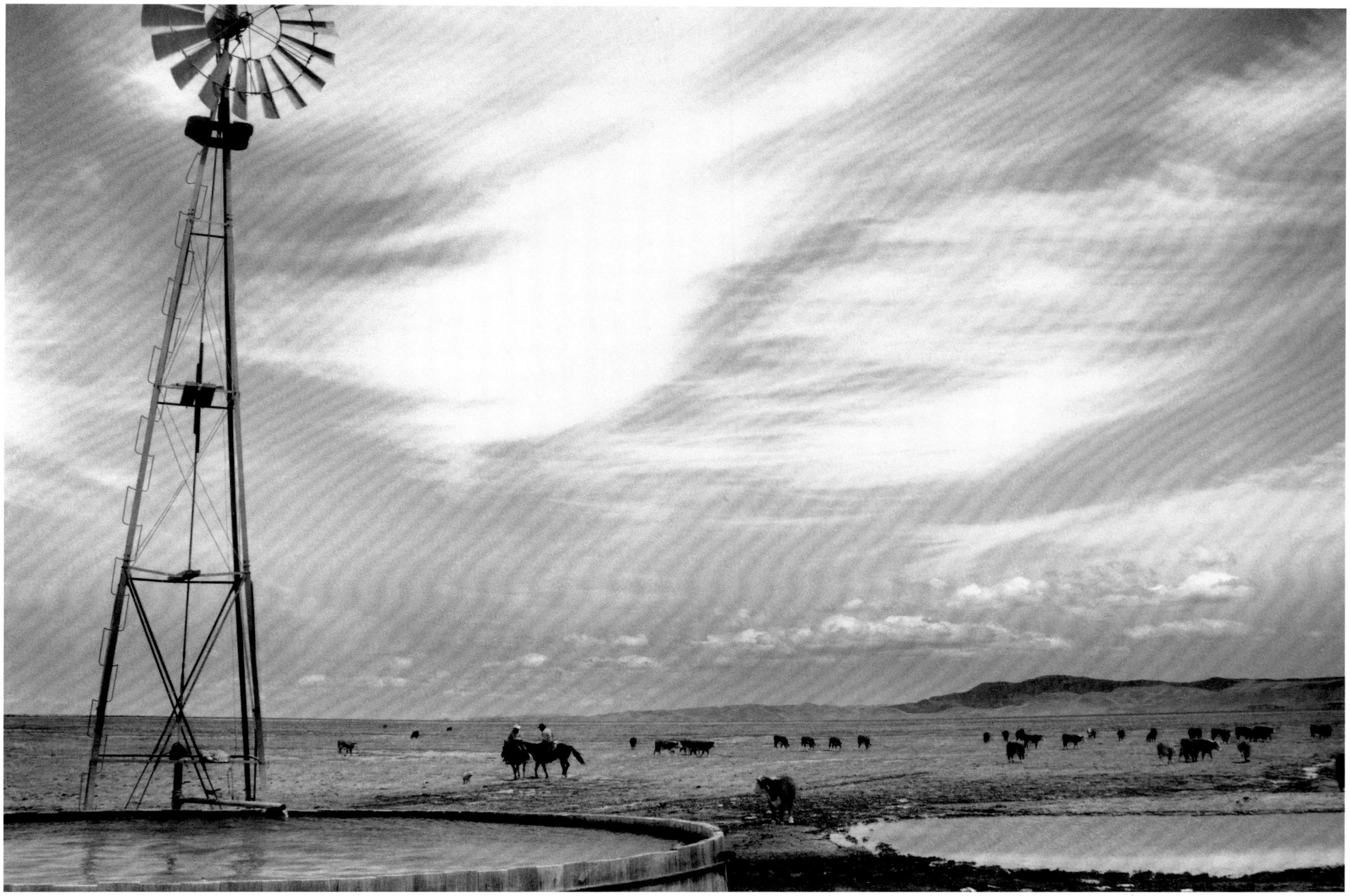

FARR CATTLE COMPANY
1988

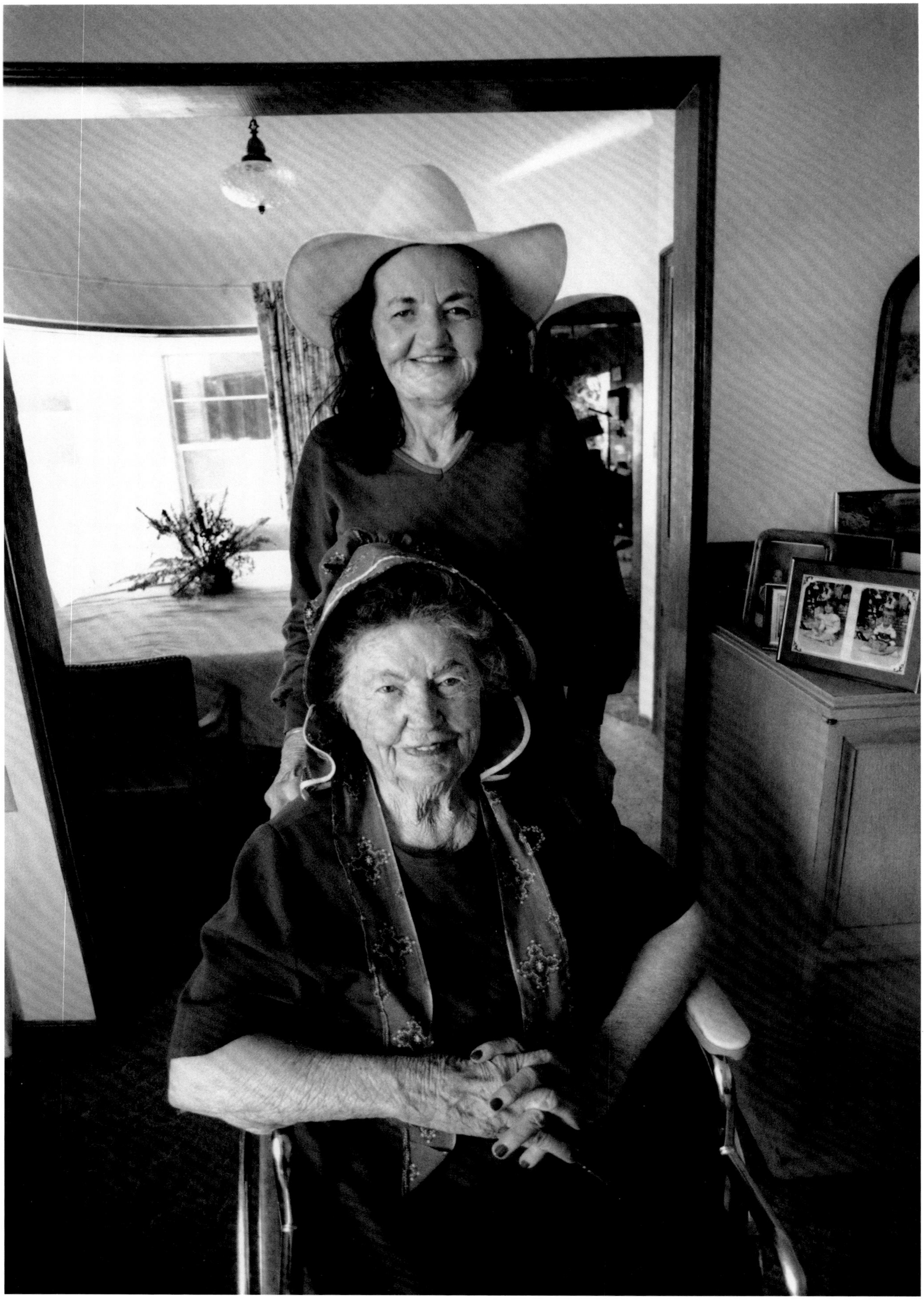

JEWELL MCADOO LUTICH WITH HER NINETY-ONE-YEAR-OLD MOTHER,
MCADOO RANCH
1988

Jewell McAdoo Lutich

McAdoo Ranch, Sierra Blanca, Texas

Jewell's family ranch lies in the heart of the great West Texas country of cattle drive history. The rolling grasslands coast up to the edges of some spectacular cliffs and deep canyons and wash up against some equally beautiful bluffs. The mountains of that country are worn down, but they provide a welcome relief to the westward-bound traveler's eyes accustomed to the flat horizon of the west-central Texas landscape. Her family ranch house is in a pretty setting—it had snowed the night before I arrived—and is large and light-filled.

Jewell, a tall, slender woman of seventy years, offers a friendly welcome, unconcerned about the considerable facial scarring that resulted from a serious car wreck that "left my face and all my teeth on the dashboard." Serendipitously, while in Galveston for plastic surgery she became involved in a lucrative sideline:

I'm one that always has had to have a job or work so I got into buying cattle at a time when there were absolutely no women doing that sort of thing. I bought cattle for the feedlots for over a year. And I made a lot more at that than I make here at the ranch. And I got a real good reputation as a cattle buyer. But then Daddy asked me to come home and run the ranch, so I've been here ever since.

Jewell was married when she was nineteen, had two children, and ran the farm that her husband and she had while he was away in the service. After the war he came back, Jewell taught school for a while, and continued helping on her daddy's ranch. The couple later divorced, she became a cattle buyer, and then moved permanently back to the ranch.

That old house with just two rooms—it's still standing—was where I was raised. We'd catch water in the dirt tank, dip it out into another tank which the mules pulled back to the house, then we'd drain that water—through a couple of towels to clean it—into a cistern, then we'd dip it out of a bucket, haul it to the house to use for drinking and cooking water. I bet no one thinks of anything like that when they turn on a faucet in a house today. All our clothes washing was done outside. We had the wash pot boiling on a campfire, poured it into the tubs, and scrubbed the clothes on those old metal scrub boards. I sure like what we have today for washing clothes. And we churned all our own butter. Believe me I know what work is! I remember when I lived over there on the old ranch and we didn't even have a light, just a kerosene lamp. Then Daddy got a wind charger and put a little globe in each room, a ten-watt globe so we had more convenient light at night. My sister and I called our "bedroom" that place under the piñon tree. We loved to sit outside because there wasn't a whole lot of room in that house. I have always liked to sleep outside.

Jewell calls herself "a wildlife cowgirl" because half of her ranch income is from hunting permits on the ranch. "Our reputation for hunting is really outstanding here. We've got lots of game." Her father started building up the ranch wildlife in 1910, and she has continued the work. She is cultivating the desert bighorn sheep, blue quail, desert mule deer, and antelope.

I run the hunting just like I would a herd of cattle. When we don't have a good production of deer, I'll cut down on the hunters. If I had to I would sell down my cattle to allow better range for the wildlife. The deer don't eat the same as cattle. They eat browse [shrubs, twigs, and leaves], but the antelope eat the same grass as cattle.

Jewell is unusually environmentally conscious. She does not use any commercial pesticides or poisons of any kind, nor does she permit them being used on the ranch.

We raise Herefords and use registered Angus bulls. I like the first cross between Hereford cows and registered Angus bulls. I think it's the best cross there is. I get my replacement heifers by using a Hereford bull on some of my best Hereford cows.

We run our bulls year-round and because our cows are so quick to "breed back" [cycle in again and be ready to be bred]—about one month after they've calved—we get the equivalent of a calf and a quarter from each cow every year. That does add up, makes that much more beef.

Because the last fifteen or so years have been dry on and off, Jewell has been feeding cottonseed cake. In 1987 alone she bought forty-eight tons of cake for her cattle. How much rainfall the ranch gets determines how much cake she needs to buy each year: "I'm beginning to think it's forgotten how to give us moisture. Droughts are tough. That will separate a rancher from a dude. You know what makes a good rancher? Plenty of rain! And you know what I'd do if I had more money than I needed? I'd figure out how to make rain!"

Jewell lives a good life. "Ranching women work 365 days a year, around the clock, and there's no '8 to 5' jobs out here. One day you're pulling a calf and you're a doctor; the next day you're fixing a well; the next day you're fencing; and the next day you're rounding up. There's never a time you 'get through' on a ranch."

She concludes, "So you can see why I say I never get lonesome—not one time have I ever. If I'm not out there, I'm in here trying to catch up!"

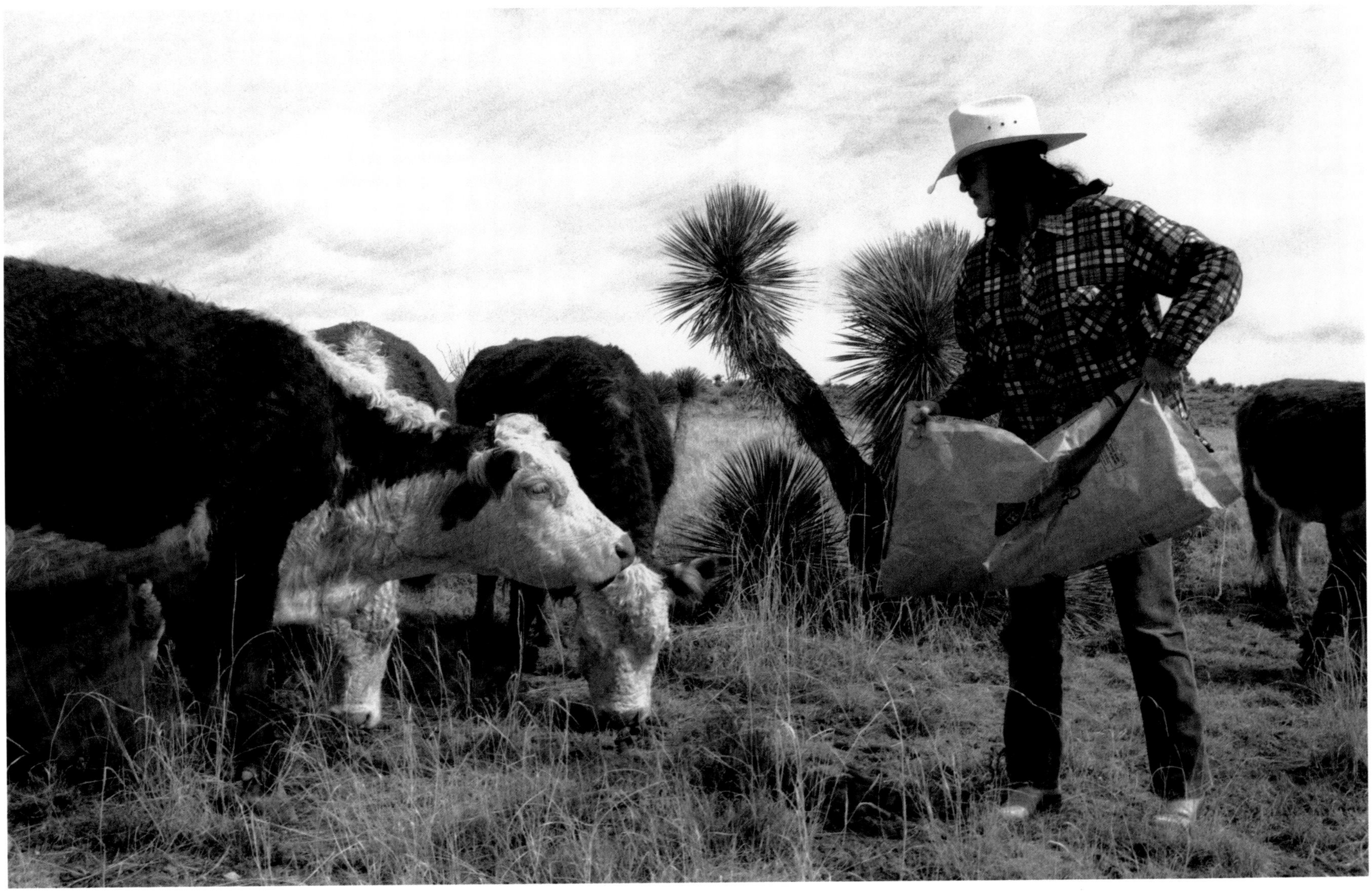

JEWELL MCADOO LUTICH, MCADOO RANCH
1988

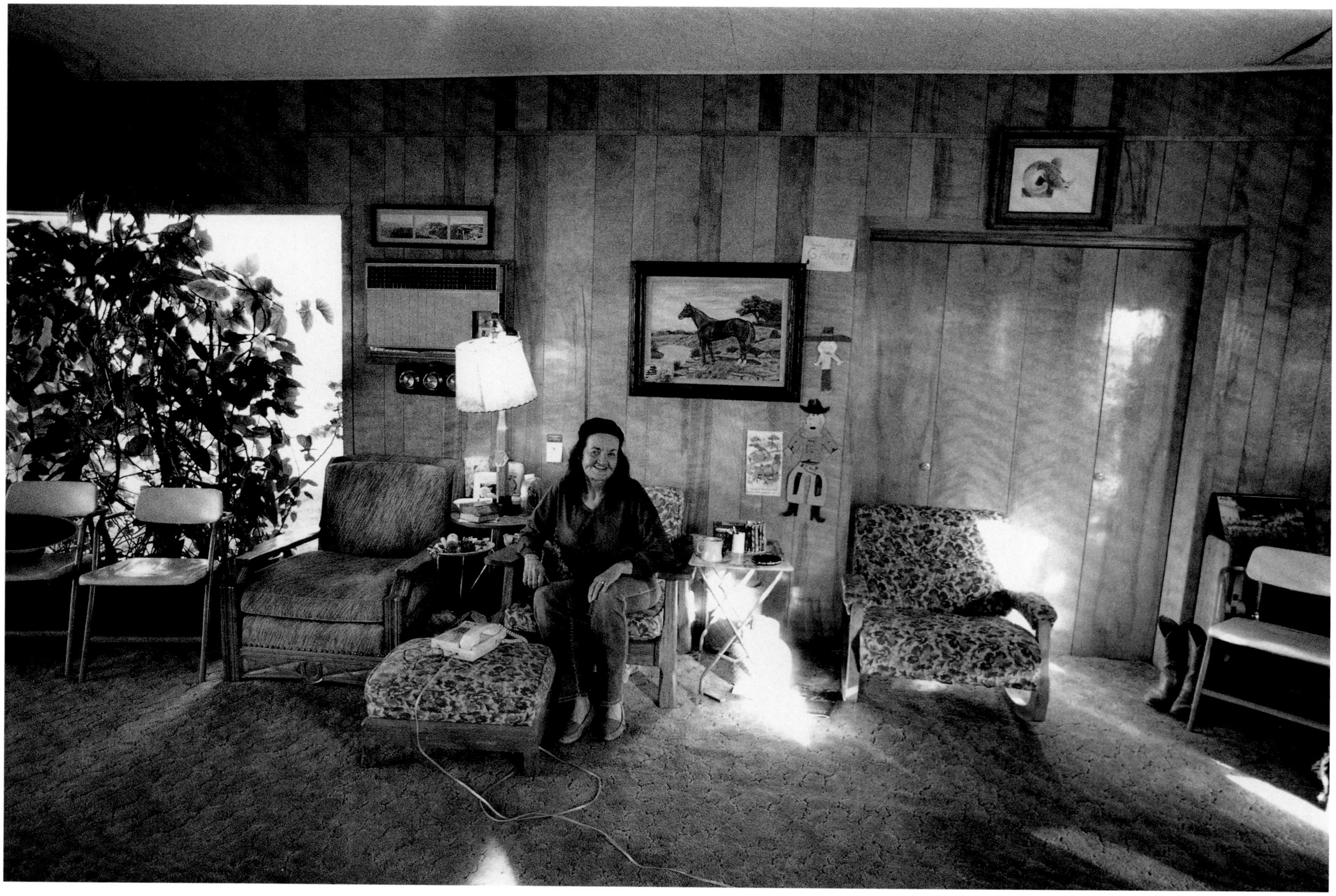

JEWELL MCADOO LUTICH, MCADOO RANCH
1988

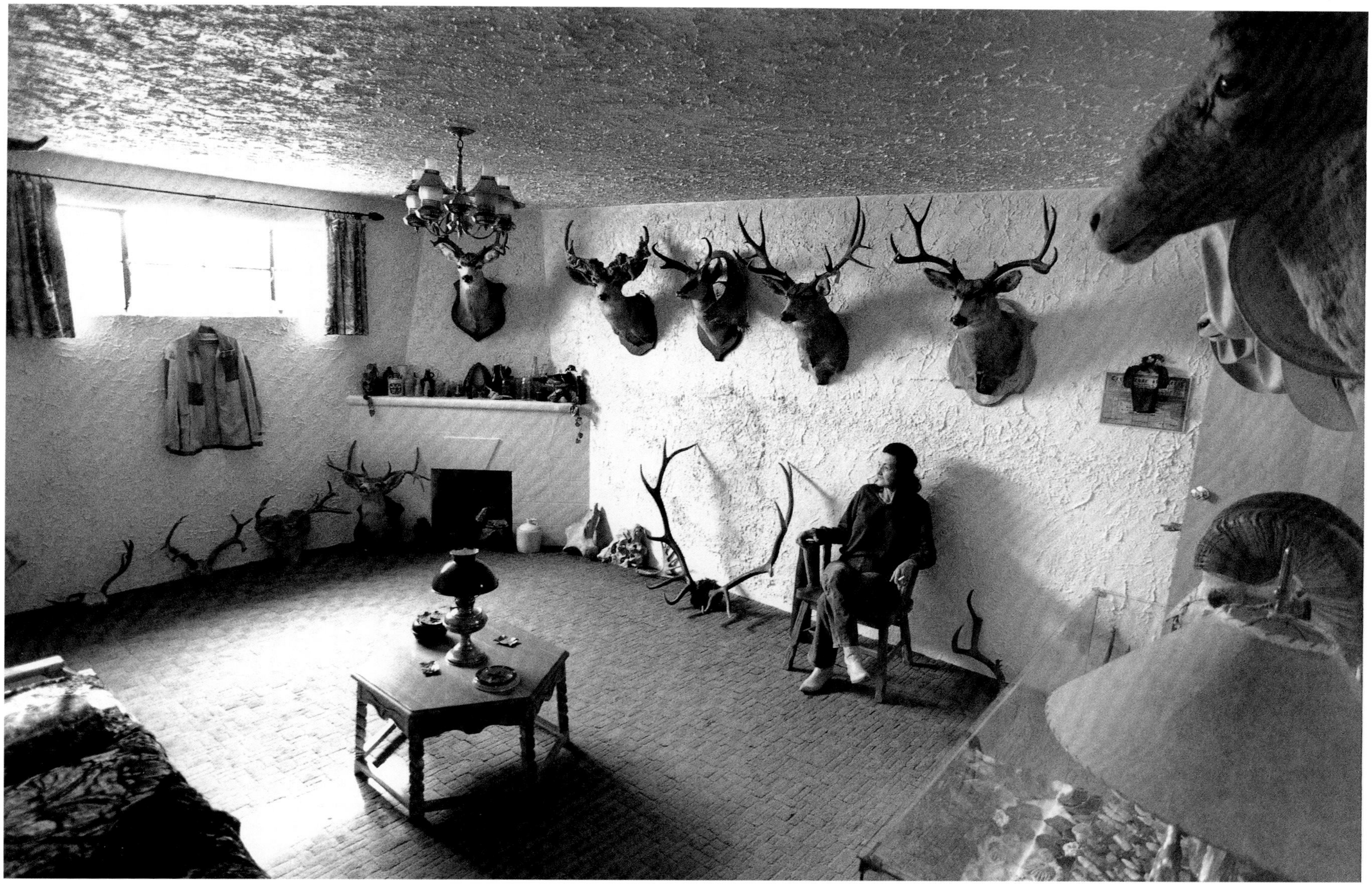

JEWELL MCADOO LUTICH, IN HER FATHER'S TROPHY ROOM, MCADOO RANCH
1988

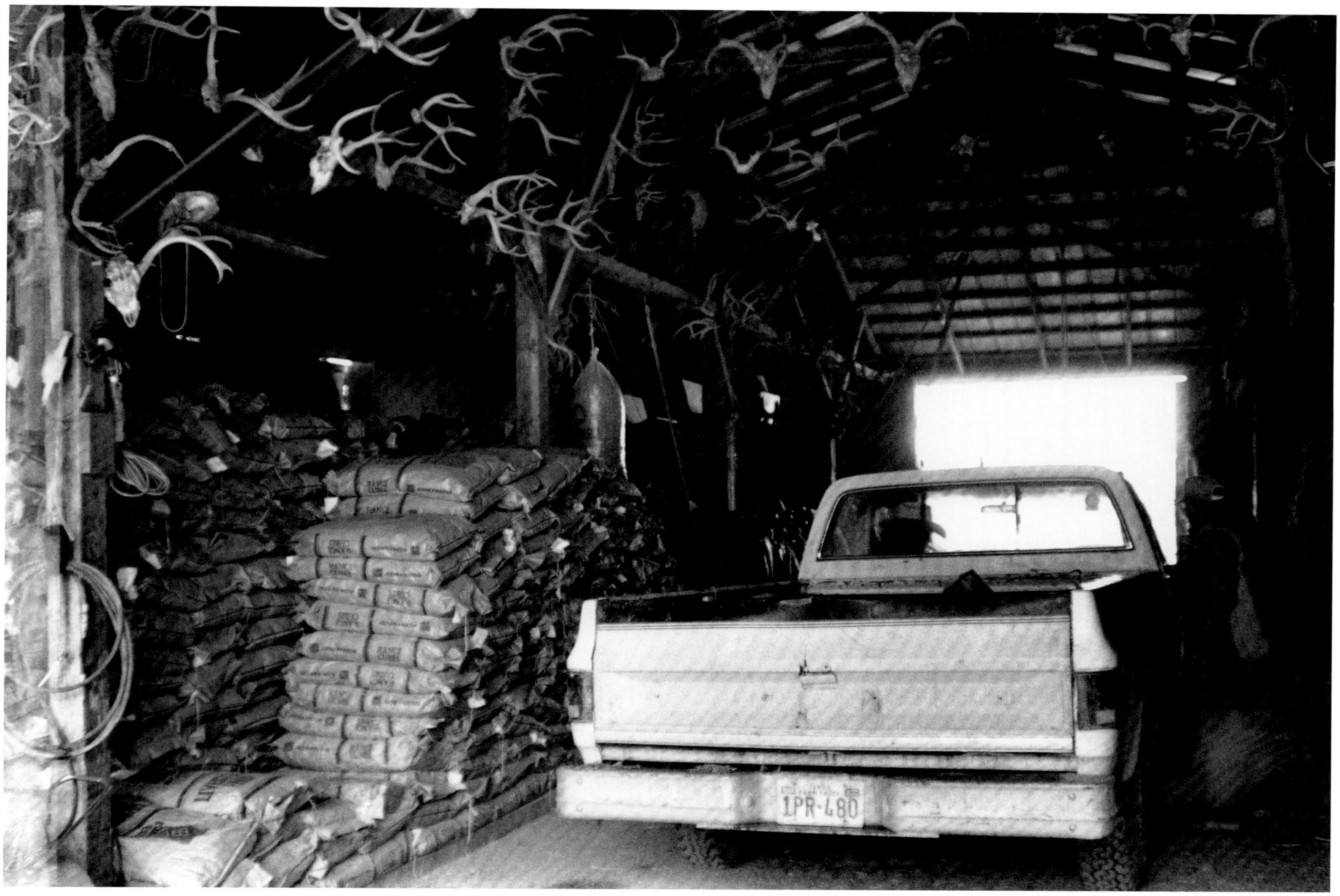

JEWELL MCADOO LUTICH, MCADOO RANCH
1988

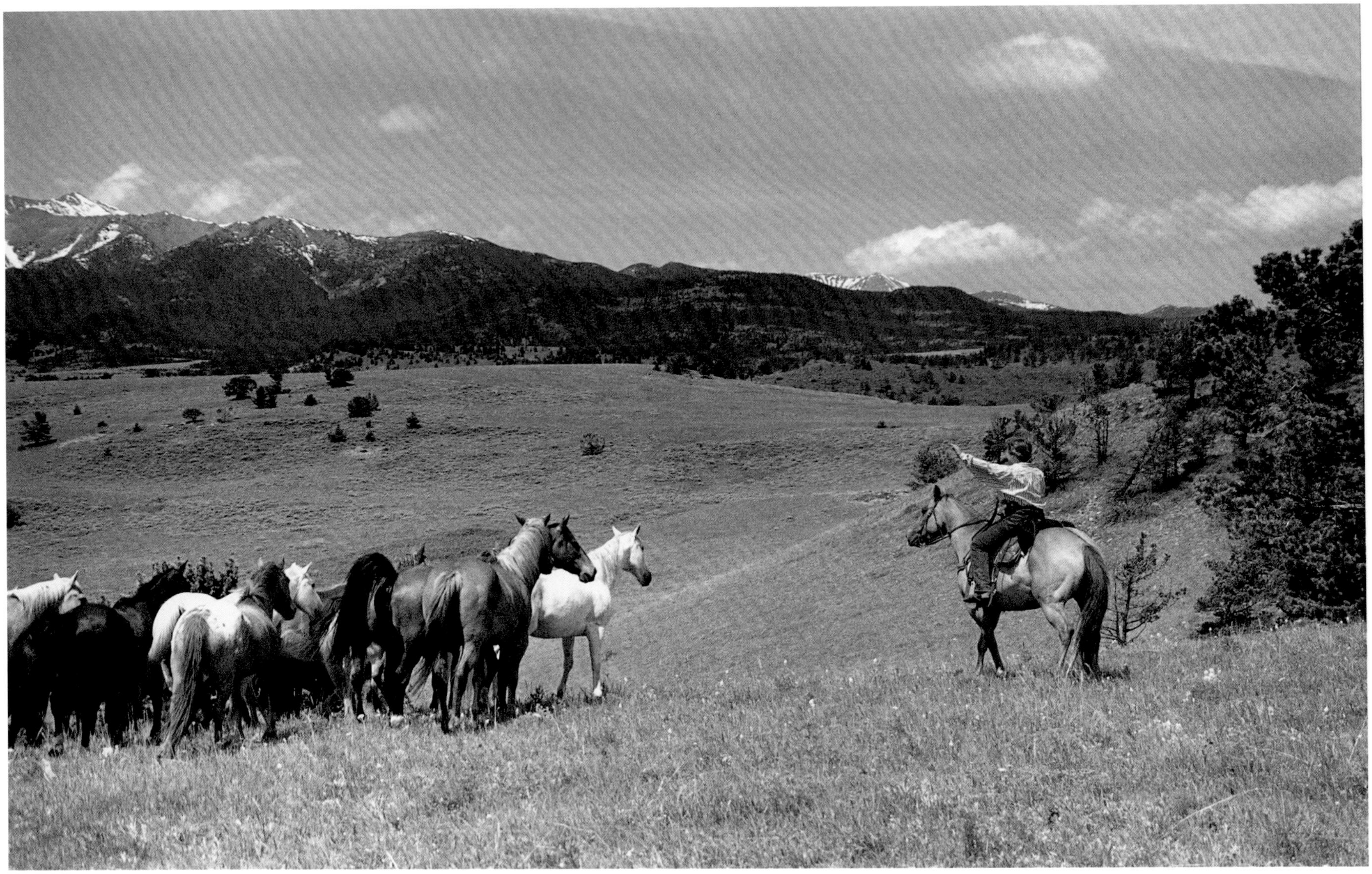

MOVING HORSES ON THE FAMILY RANCH, OUTSIDE BIG TIMBER, MONTANA
1995

AFTERWORD

Barbara Van Cleve's view of modern ranch life is a refreshing, clear-eyed look at a subject that has long been romanticized to a point where it has only a tenuous connection with reality. Van Cleve's ranches are peopled with real women and men who labor at the often arduous task of wrenching a living from a beautiful but hard land. And for one of the first times, the primary focus of this portrait of the modern West is the women who own these ranches, work on them, preserve them, and manage their precious resources. Here, at last, after a century of portraying the West as almost wholly the domain of men, is a realistic picture of women living and working as equal partners with men in the running of ranches.

In fact, these women may be more than equal since their obligations so often include all of the domestic duties of cooking and running a household along with many of the tasks of working cattle as well. While each of the women portrayed here and each of the ranches chronicled are steeped in tradition and committed to maintaining a lifestyle that has persisted over the course of a century and more, these photographs also show women who are adaptive to modern methods of ranching.

What is evident in these photographs is that in spite of the hard work, in spite of the uncertainty of scratching out a living from an uneasy truce between land and cattle, in spite of the vicissitudes of an economy that they have little or no control over, these women are survivors who love the life they are living. They belong to the places pictured here and are connected to and rooted in the day-in and day-out rhythm of the ranch. Some have traveled widely, have been educated in the nation's finest universities; others have stayed close to home and been educated by experience; but all are grounded in the work of running a ranch and tuned to the natural cycles of seasons or climate, of the primordial connection between humans and animals, and the land.

Each of these ranch women plays a unique role, but each is part of the larger tradition of western living. These women have been shaped by their past experiences while helping to shape a new generation who will take over the reins of maintaining a traditional ranch operation. Preserving that lifestyle is of prime concern to these women and will surely test their ingenuity and creativity. The economics of running such family operations get more difficult with each passing year. Small families have trouble simply manning the operation while larger ones risk the dissolution of the ranch through division into separate parcels. Yet in spite of these difficulties, there is no indication that the traditions these women are attempting to preserve will fade away. Their strength, determination, and humor are apparent in these images. What is shown here is a resolute spirit, one that will see these women through the changes, good and bad, to come.

Pictures of the West by artists, photographers, and writers beginning in the nineteenth century and continuing in some measure today are largely devoid of the actual work that cattle ranching involves. The figure of the lone cowboy loping across the plains is a familiar one to all of us. It has been largely through these pictorial representations that the cowboy has become something of an American icon, adopted by artists and writers alike as a mythic hero and used by the advertising industry to sell everything from cigarettes to cologne. The result of this one hundred years of legend making is that we have come to accept the cowboy in a number of guises—the gun fighter, the lawman, the avenging stranger (such as Shane or any of a number of Clint Eastwood heroes), but not as someone who works with cattle (with a few notable exceptions). One of the most famous cowboy novels, *The Virginian*, by Owen Wister, has a main character who is a ranch foreman who never comes close to a branding iron. However, the laconic Virginian captured the American imagination like few other figures before him and has held on with the tenacity that any modern bull rider would admire.

That novel also invoked a female figure that proved to be popular as well, at least with those in the business of creating western fictions. The heroine of the novel was a young woman newly arrived from the East who knows nothing of the rough and tumble ways of the West. She helped create a whole cast of stereotypical women of the West who would be found in western paintings, novels, and even photographs for years to come. For these artists, women were most often presented as school teachers, saloon girls, Indian princesses, or the staple of western fiction, the spunky rancher's daughter—spunky only until she was put in her place by the strong, silent cowboy hero. The real women of the West, the ones who homesteaded, the ones who performed the same sort of back-breaking work as their husbands, fathers, and brothers, are rarely seen.

Instead, a classic example of that earlier view of western women was provided by artists such as W. H. "Buck" Dunton, one of the founders of the Taos Society of Artists. Dunton, who had been trained as an illustrator early in his career, often painted a sort of "Gibson Girl" of the West, shown in long flowing dresses and skirts, often with a flower-bedecked hat. She usually rode side-saddle across prairies strewn with wildflowers or was seen lounging at pleasant picnics with a few bashful cowhands nearby. For the most part, these were women who only ventured outdoors in good weather under brilliant blue western skies. They are not women who would have been comfortable mending barbed wire fences or helping a cow through a difficult birth.

Dunton's paintings and literally thousands of others like them were widely circulated in such mass-market publications as the *Saturday Evening Post.* Generations of Americans grew up looking at these pictures and reading the accompanying stories that were purportedly drawn from first-hand observations. Later, these images were literally brought to life through the magic of the moving picture. While Hollywood is often blamed for creating myths, in this case screenwriters and directors were only following in the well-worn paths of the illustrators. The cowboy became a staple of American culture, and his feminine companions, if he had any at all, were kept firmly in the background.

Such an overwhelmingly masculine image of the West was not confined just to painting and fiction. Nineteenth- and early-twentieth-century photographers who chronicled ranch life were equally guilty of literally leaving women out of the picture. L. A. Huffman's evocative images of Montana cowboys do show a fairly realistic picture of early ranch life, but from the point of view of the cowboy. Other photographers such as Ray Rector in Texas and Charles Belden in Wyoming and Montana set out to document the life of the range cowboy and did so while largely ignoring any contributions to range life by women. Irwin Smith, whose photographs of Texas ranch life are probably the most famous of that genre, used his training as a sculptor to arrange his cowboys in group poses and to create dramatic images of cowboy life that were generally true to the spirit of that life, if not to actual details. Smith, however, was not averse to a little artistic license, such as in his widely published photograph of a cowboy pulling a letter from a mailbox perched on a fence post in the middle of the pasture. That picture is a direct copy of an earlier painting by illustrator Rufus Zogbaum, which was also copied by artist N. C. Wyeth.

Together the work of these photographers and that of their contemporary artists and writers presents a view of the West that is pretty far removed from reality. It is a land where women simply are absent. That masculine image still persists today in the works of contemporary painters and photographers, but with a few new twists. Today, the romantic nature of picturing the West often emphasizes the end of an era, the twilight of cowboy life. Numerous photographic portraits of the West and ranching have been published over the last twenty years or so which have lumped cowboys in the category of the "vanishing breed." Many writers have also attempted to document the "last cowboy." Western art, always romantic, has developed a decidedly nostalgic vein when it comes to depicting cowboys and ranching. Ironically, the figure of the cowboy, which in the past stood for a certain American vigor and sense of adventure, has increasingly come to be seen as representative of the past or, at the very least, of the fading present. In this sentimental view, the cowboy has become a figure both out of place and out of time, with actually very little connection to modern ranching.

Of course there are many historians and critics alike who would say that the romantic figure of the cowboy has very little to do with an accurate portrayal of historic ranching. Certainly, as many new histories of the West (often written by women) tell us, women were not absent from the historic West. They have been there all along, engaged in the business of raising cattle and horses and the myriad tasks of establishing homes and keeping family operations going. The stories of many of the women in this book stem directly from their mothers and grandmothers who were at home both on the range and in the parlor. Occasionally the faces of these women can be seen in photographs that are scattered throughout historical societies, museums, libraries, and universities. They are often anonymous images of women at work in a variety of tasks that do not lend themselves easily to sentimental renditions of bucolic ranch life, but these are real people and their stories more than likely contain enough drama and heroism to fill dozens of novels by Zane Grey or to produce hundreds of paintings by western artists.

These are the parents and grandparents of the women that Barbara Van Cleve has chosen to portray, women who exemplify the modern West and who no doubt will have much influence on the shape of the future West. Their stories lie in the details of these photographs, the often mundane acts of daily living that can change in the blink of an eye to become extraordinary experiences. Work is one of the primary focuses of this study and it is in these photographs that Van Cleve's insight into the real life of the West is so apparent and such a sharp contrast with earlier and present romantic portraits. These women are tied to the land they live on and to the animals that they raise, and their lives are regulated by the necessities of those two elements.

Van Cleve takes note of the ordinary and turns those moments into images of both beauty and drama. Her link to these women is one borne of her own experiences and her own connection to the life she is portraying. And that may explain in some ways why so much of the imagery of the West has been lacking in this hard dose of reality. In many cases, it has been produced by men who may have been keen observers but in the end were only passing through. Their connection to the subject was transitory. Van Cleve's connection is from the inside, from someone who has ridden down the same road and who is as connected to the traditions of family ranching as the women in these photographs are.

What emerges in the pages of this book is a spirit that is far removed from much of the popular conception of the West. There are scenes here of stunning beauty that are as breathtaking as any of the canvases of Albert Bierstadt and Thomas Moran; there are images of high adventure and drama that rival those of the best of contemporary western painters, but there is also more. In these women, we can see both the present and the future of the West. These women are not members of a "vanishing breed," nor reminders of the good old days that are good and gone. These are women who are going to get up tomorrow, and the next day, and the day after that to get back to those jobs that may not be the subjects of western art exhibitions, that may not make the pages of another glossy look at the "dying" cowboy, but which are jobs that having been done well will assure that the daughters and sons of these women will also have a deep connection to a particular place and kind of life.

It is a life that is often hard, but there is no evidence in the faces of these women that it is unsatisfying. With these photographs, we get a glimpse of that life and those feelings and we are able to appreciate the nature of working a ranch and seeing the real West without the romantic haze that may have created heroic figures but obscured the true heroines and heroes.

—Michael W. Duty
Walnut Creek, California

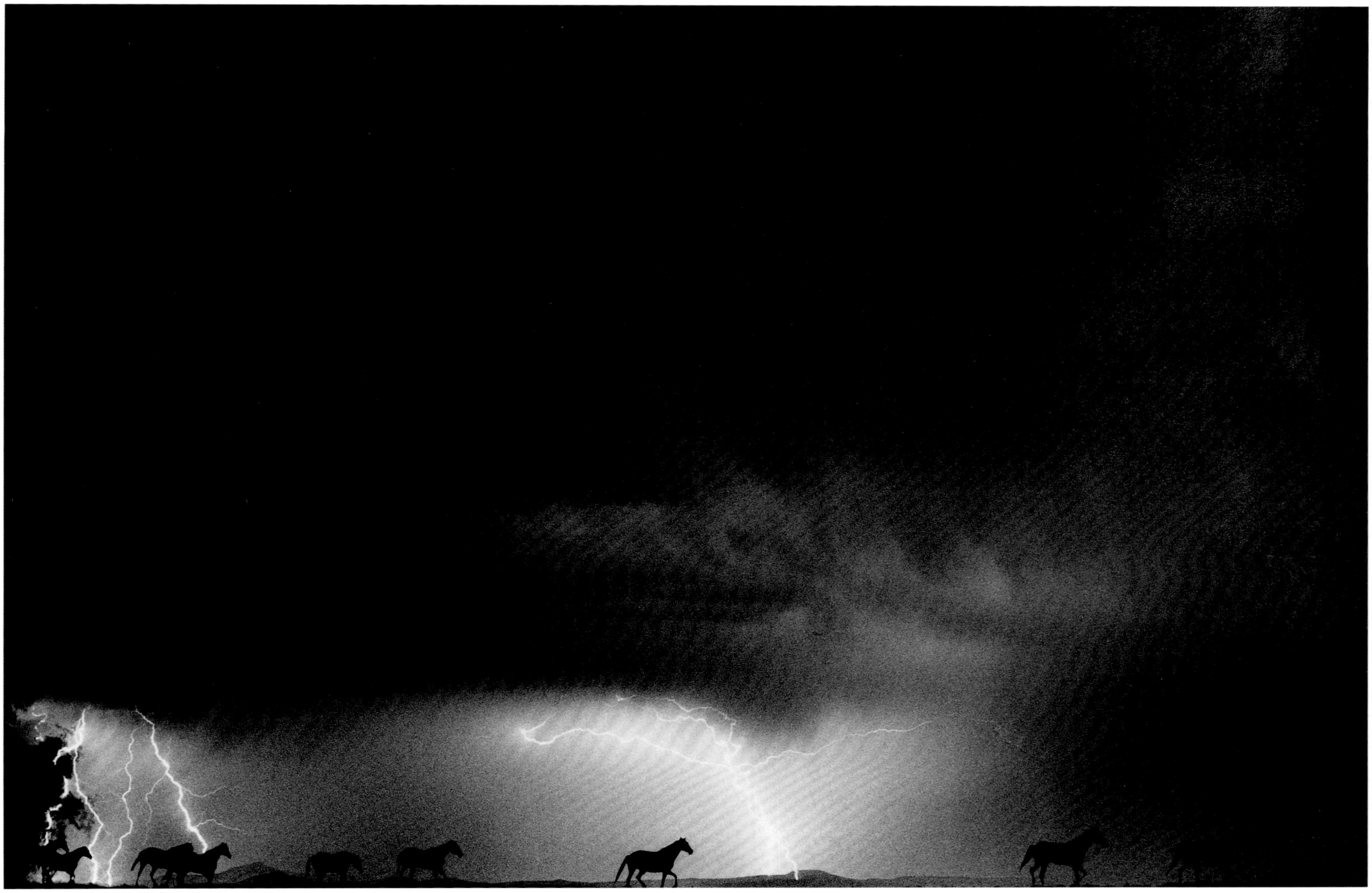

HOT STORM
1993

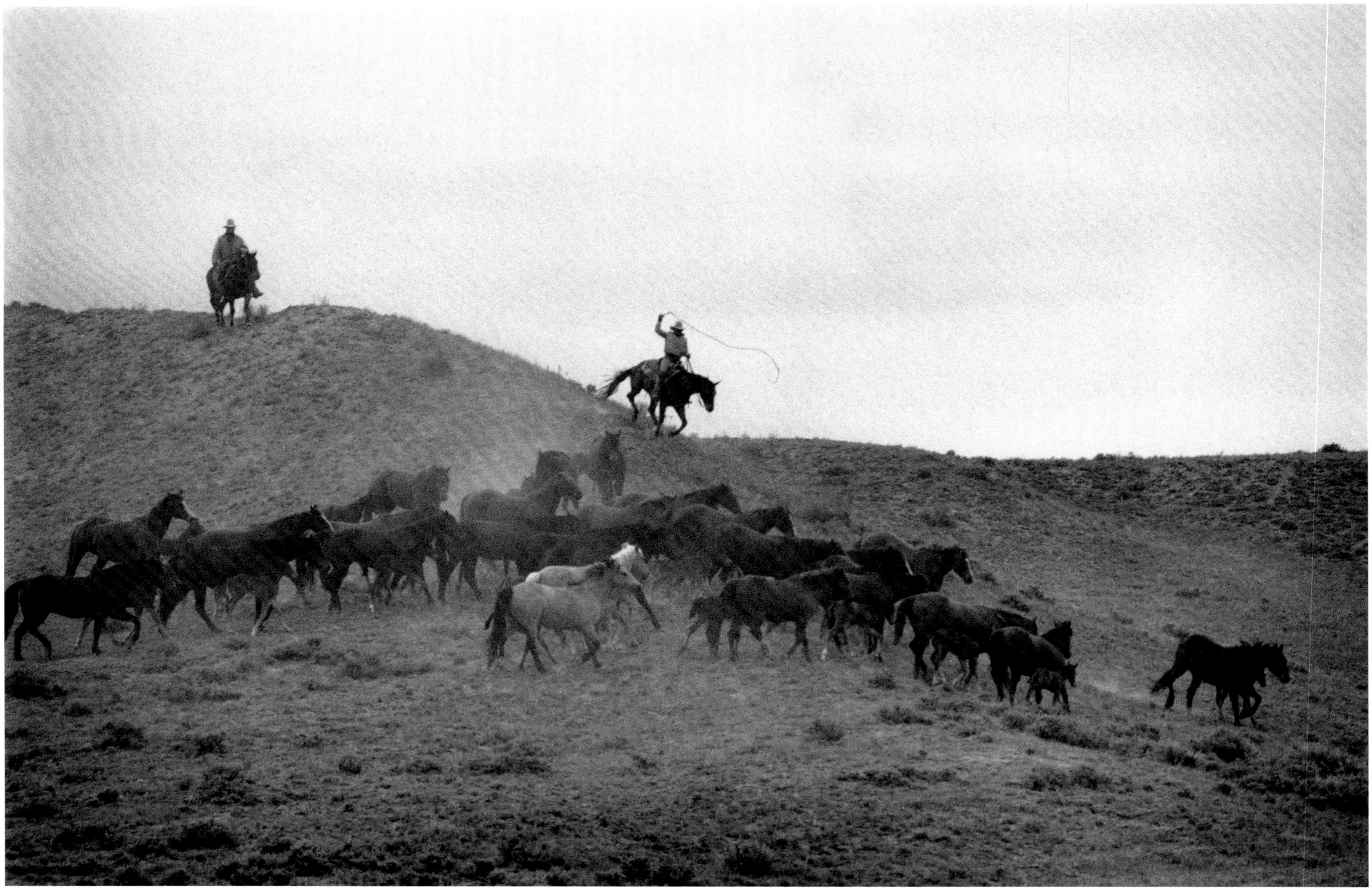

DALE AND CARLA KAMP, SCOTT HORSE RANCH, BILLINGS, MONTANA
1987

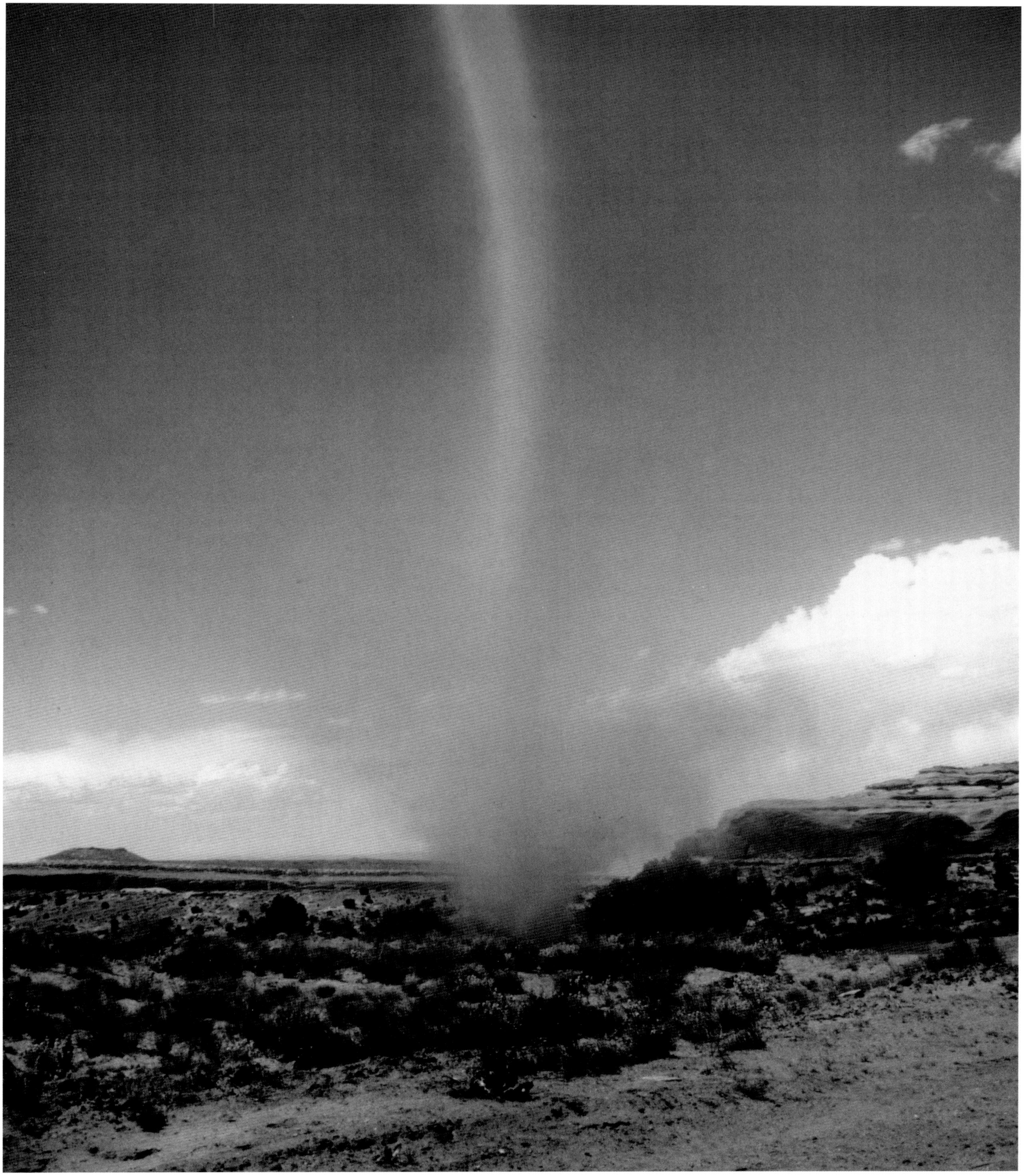

DUST DEVIL, UTAH
1993

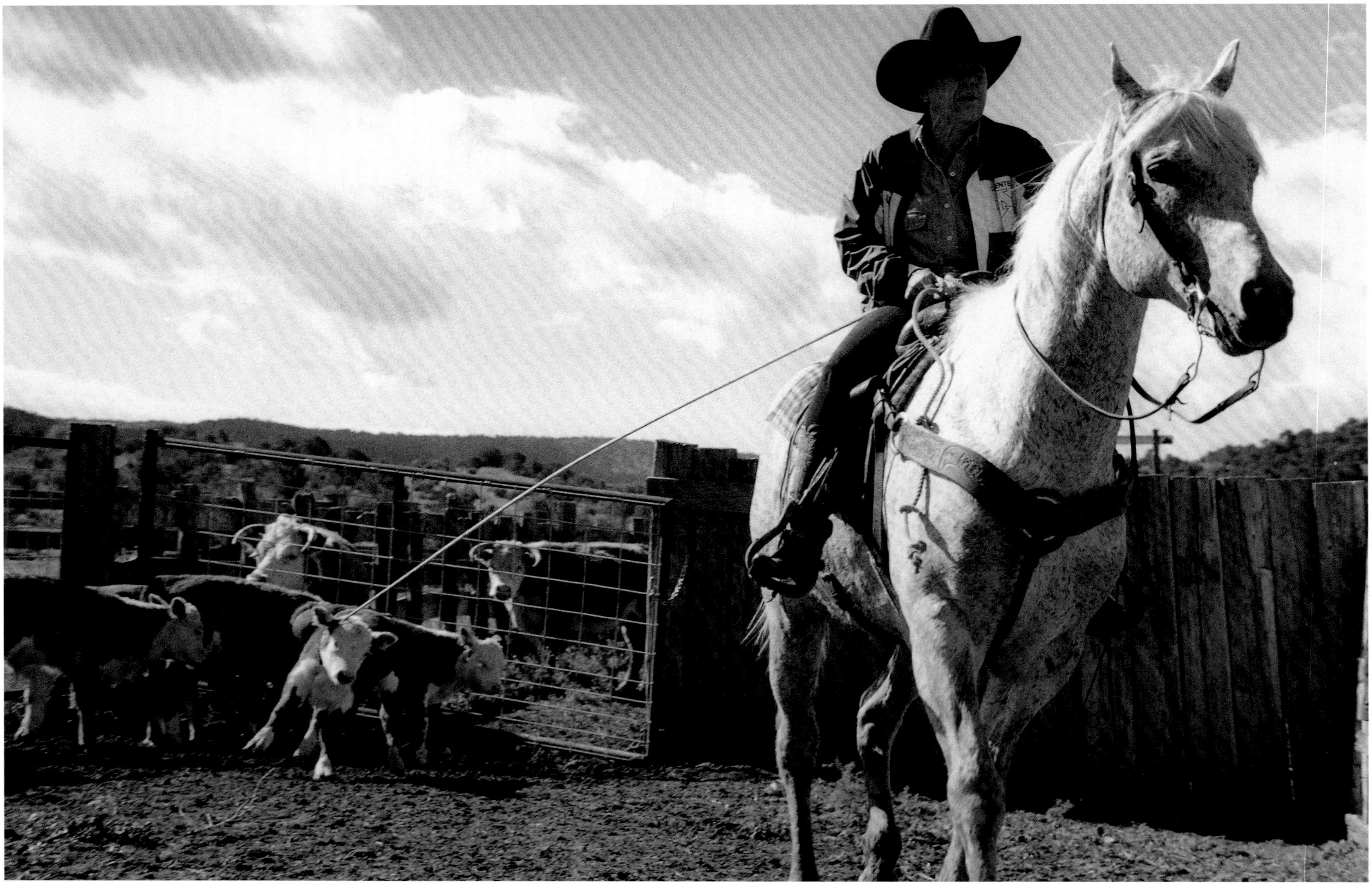

FERN SAWYER, RICHUELA RANCH, NOGAL, NEW MEXICO
1987

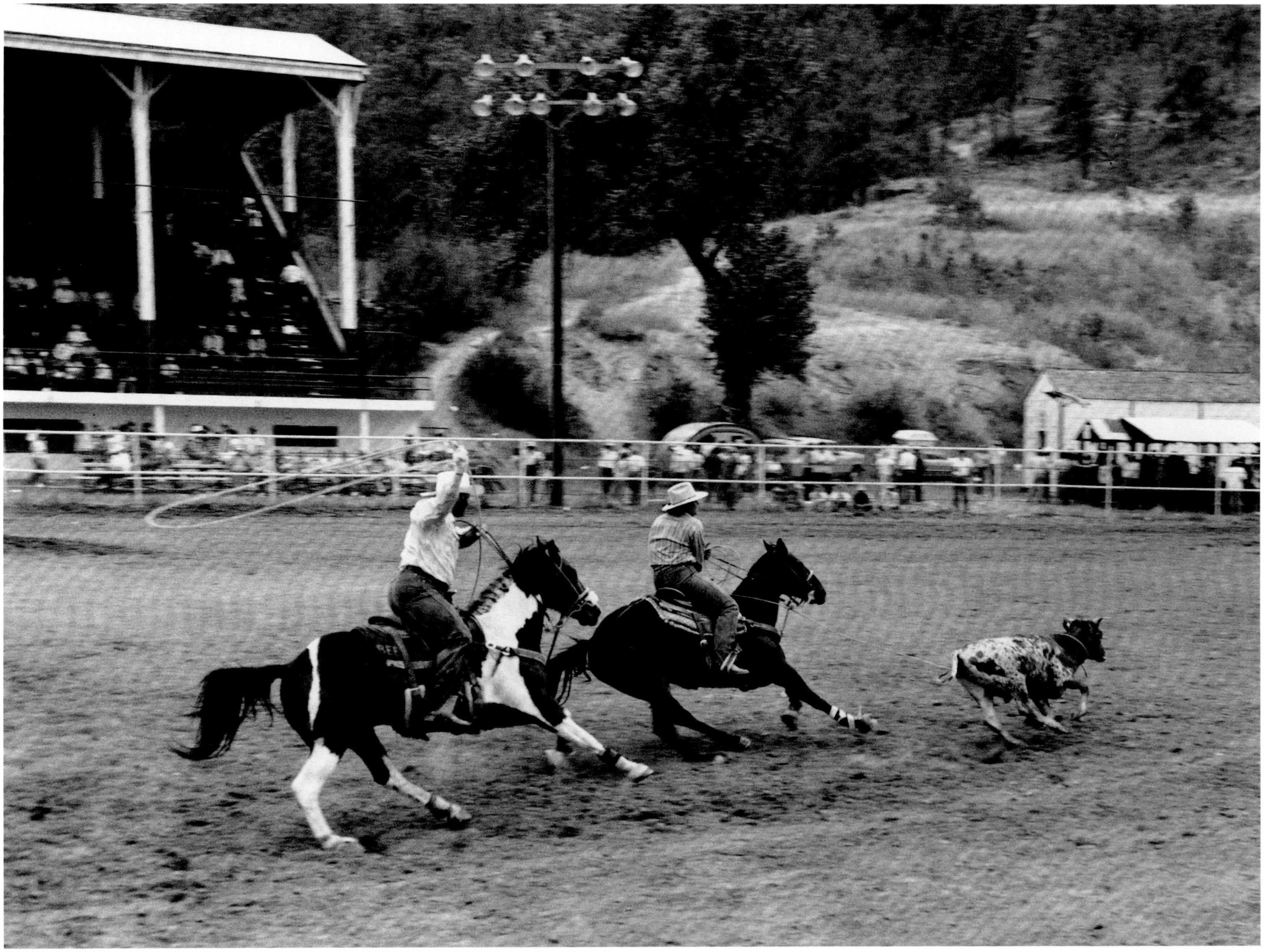

DAVID AND JANIE CARLSON, TEAM ROPING AT RYEGATE, MONTANA, RODEO
1986

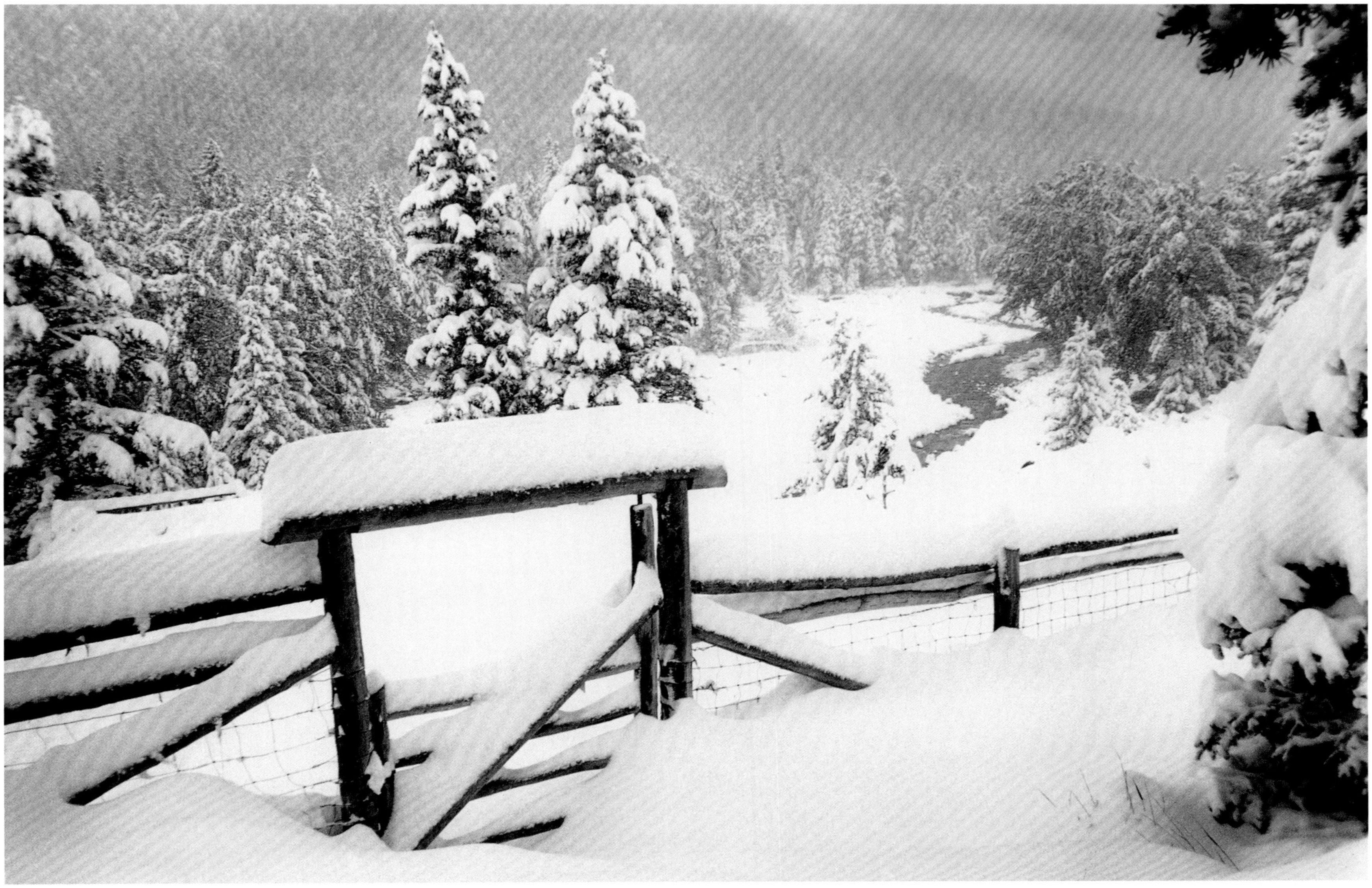

WINTER'S PREMIERE, "AT THE HEAD OF THE CREEK"
1982

ABOUT THE AUTHOR

Barbara Van Cleve's heritage is rich with family history and firsthand experience. Her family's ranch, the Lazy K Bar, was founded in 1880 on the east slopes of the Crazy Mountains near Melville, Montana. Her father, Spike Van Cleve, was a unique combination of writer, poet, rancher, and expert horseman.

Her Montana is a lifetime of knowing the hard work of ranching beneath the spectacular slopes of the Crazy Mountains. Although she photographs landscapes, portraits, and documentary subjects, it is her photography of the ranching West that has earned her a national reputation. She is collected internationally as well.

Barbara Van Cleve is the author, with cowboy poet Paul Zarzyski, of *Roughstock Sonnets*. In 1995 she was inducted into the National Cowgirl Hall of Fame. In 1996 a major traveling exhibition, "Hard Twist: Western Ranch Women," opens in Oklahoma City at the National Cowboy Hall of Fame.

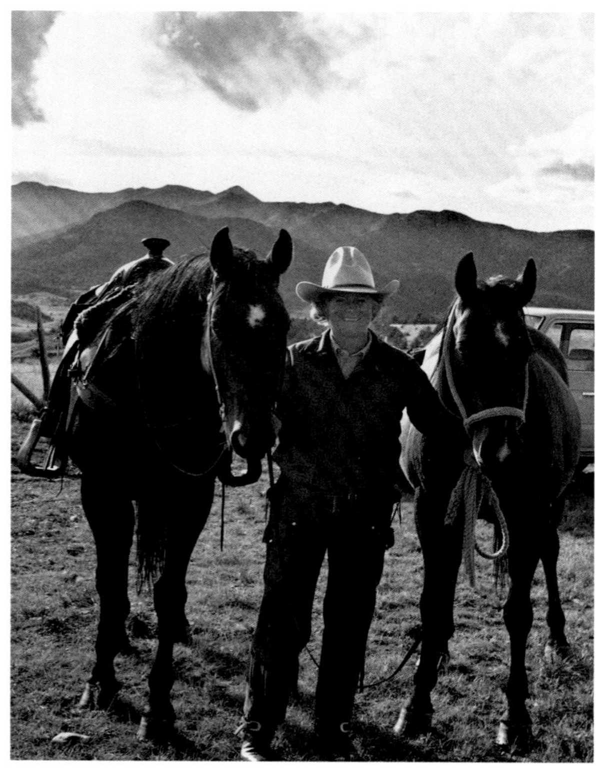

BARBARA VAN CLEVE WITH SHIZZERS (LEFT) AND HVERFANA, 1989. PHOTOGRAPH BY DEEDE PHILLIPS.